Clinical Sciences Review for Medical Licensure
Developed at
The University of Oklahoma College of Medicine

Ronald S. Krug, *Series Editor*

Suitable Review for:
**United States Medical Licensing Examination
(USMLE), Step 2**

**Springer**
*New York
Berlin
Heidelberg
Barcelona
Budapest
Hong Kong
London
Milan
Paris
Santa Clara
Singapore
Tokyo*

# Neurology and Clinical Neuroscience

Second Edition

Roger A. Brumback

Rita R. Claudet
Technical Editor

Springer

Roger A. Brumback, M.D.
Departments of Pathology and Neurology
University of Oklahoma College of Medicine
Oklahoma City, OK 73190 USA

Library of Congress Cataloging-in-Publication Data
Brumback, Roger A.
    Neurology and clinical neuroscience / Roger A. Brumback :
technical editor, Rita R. Claudet. — 2nd ed.
      p.   cm. — (Oklahoma notes)
    Includes bibliographical references.
    ISBN 0-387-94635-7 (softcover : alk. paper)
    1. Neurology—Outlines, syllabi, etc.   2. Neurology—Examinations,
questions, etc.   I. Title.   II. Series.
    [DNLM: 1. Nervous System Diseases—outlines.   2. Nervous System
Diseases—examination questions.   WL 18.2 B893na 1996]
RC357.B78   1996
616.8'076—dc20
DNLM/DLC
for Library of Congress                                             96-13784

Printed on acid-free paper.

© 1996, 1993 Springer-Verlag New York, Inc.
All rights reserved. This work may not be translated or copied in whole or in part without the written permission of the publisher (Springer-Verlag New York, Inc., 175 Fifth Avenue, New York, NY 10010, USA), except for brief excerpts in connection with reviews or scholarly analysis. Use in connection with any form of information storage and retrieval, electronic adaptation, computer software, or by similar or dissimilar methodology now known or hereafter developed is forbidden.
The use of general descriptive names, trade names, trademarks, etc., in this publication, even if the former are not especially identified, is not to be taken as a sign that such names, as understood by the Trade Marks and Merchandise Marks Act, may accordingly be used freely by anyone.
While the advice and information in this book is believed to be true and accurate at the date of going to press, neither the authors nor the editors nor the publisher can accept any legal responsibility for any errors or omissions that may be made. The publisher makes no warranty, express or implied, with respect to the material contained herein.

Production managed by Robert Wexler; manufacturing supervised by Joe Quatela.
Camera-ready copy prepared by the author.
Printed and bound by Edwards Brothers, Inc., Ann Arbor, MI.
Printed in the United States of America.

9 8 7 6 5 4 3 2 1

ISBN 0-387-94635-7 Springer-Verlag New York Berlin Heidelberg  SPIN 10522826

# Preface to the Oklahoma Notes

The intent of the Oklahoma Notes is to provide students with a set of texts that present the basic information of the general medical school curriculum in such a manner that the content is clear, concise and can be readily absorbed.

The basic outline format that has made the Oklahoma Notes extremely popular when preparing for standardized examinations has been retained in all the texts. The educational goals for these materials are first to help organize thinking about given categories of information, and second, to present the information in a format that assists in learning. The information that students retain best is that which has been repeated often, and has been actively recalled. The outline format has always been used in the Oklahoma Notes because students have reported to us that it allows them to cover subsequent parts of the outline, and use the topic heading as a trigger to recall the information under the heading. They then can uncover the material and ascertain whether they have recalled the information correctly.

This second edition of the Clinical Series of the Oklahoma Notes represents a major refinement of the first edition. A number of issues have been addressed to make the texts more efficient, effective and "user friendly." These include:

- Correction of technical errors.
- Addition of new material that has been reported since the first editions were published.
- Standard presentation of materials in all texts to make information more accessible to the student.
- Review questions written in standardized format. These questions reflect the major issues of the sections of the texts.

We hope these are helpful to you in your educational progress and preparation for required examinations.

Ronald S. Krug, Ph.D.
*Series Editor*

# Preface

The United States Congress has designated the 1990s as the "Decade of the Brain" in recognition of the major importance of neurology and the other neurosciences in the health and well-being of Americans. It has been suggested that as many as 20% of all patients seeking medical treatment have neurologic problems, either as the presenting complaint or as an associated condition complicating the primary illness. Thus, it is fitting that Springer-Verlag should acknowledge the prominence of this medical specialty area by devoting an entire volume of the *Oklahoma Notes* series to neurology and clinical neuroscience.

Of course, this text is an outline overview and does not attempt to provide encyclopedic coverage of neurology (the student desiring a comprehensive review of the field may wish to seek in the library the 60+ volumes in the series *Handbook of Clinical Neurology* edited by Pierre J. Vinken and George W. Bruyn). However, the information selected for inclusion in this volume of the *Oklahoma Notes* series remains true to the goal of the whole series—only materials vital in both the general clinical practice of medicine and to answer questions on the all-important United States Medical Licensing Examination have been incorporated in the text.

Roger A. Brumback, M.D.
Oklahoma City

# Acknowledgments

This text would not have been possible without a great deal of help and support over the years from a number of individuals. I owe the fundamentals of my education in neurology and neuroscience to my mentors William M. Landau and Philip R. Dodge of the Washington University School of Medicine and Lowell W. Lapham of the University of Rochester Medical Center. These three great teachers demonstrated how a thorough knowledge of basic neuroscience could be used to understand and solve clinical neurologic problems. Richard W. Leech (former Chairman) and Fred G. Silva (current Chairman) of the Department of Pathology of the University of Oklahoma College of Medicine have provided me with a supportive environment in which to pursue my teaching and writing.

Elizabeth Claire Maletz prepared drawings for the first edition of this volume and Traci Louise Tullius provided additional drawings for the second edition. I want to dedicate this text to my wife Mary Helen Brumback, whose assistance has been vital in finalizing the volumes. This is another example of her unselfish devotion and support that has sustained me throughout my career.

Roger A. Brumback, M.D.
Oklahoma City

# Contents

| | |
|---|---|
| Preface to the *Oklahoma Notes* | v |
| Preface | vii |
| Acknowledgments | ix |

**Chapter 1  Neurologic Exam and Neurodiagnostic Tests** ......... 1
   History taking ......... 1
   Neurologic examination ......... 1
   Lumbar puncture ......... 13
   Computed tomographic (CT) scans ......... 14
   Magnetic resonance imaging (MRI) ......... 14
   Myelography ......... 15
   Cerebral angiography ......... 15
   Isotope cisternography ......... 15
   Electroencephalography ......... 15
   Electromyography (EMG) ......... 15
   Nerve conduction velocity (NCV) study ......... 16
   Audiometry ......... 16
   Evoked potentials ......... 16
   Neurosonography ......... 17
   Muscle biopsy ......... 17

**Chapter 2  Stroke** ......... 18
   Terminology ......... 18
   Lesion localization ......... 18
   Clues in diagnosis ......... 19
   Selected stroke syndromes ......... 20
   Hypertensive hemorrhagic stroke ......... 22
   Subarachnoid hemorrhage ......... 23
   Inflammatory diseases of cranial arteries ......... 24
   Sickle-cell anemia ......... 25
   Hypoxia ......... 25
   Border zone infarction ......... 25
   Aphasia and language disturbances ......... 26

**Chapter 3  Epilepsy and Seizures** ......... 28
   Terminology ......... 28
   Characteristics of seizures ......... 28
   Epileptic phenomena ......... 29
   Anticonvulsants ......... 30
   Electroencephalogram ......... 30
   Grand mal epilepsy ......... 31

Petit mal epilepsy ... 31
Benign centrotemporal epilepsy ... 32
Neonatal seizures ... 32
West syndrome ... 33
Lennox-Gastaut syndrome ... 34
Benign febrile seizures ... 35
Focal motor seizures ... 35
Psychomotor or temporal lobe seizures ... 36
Adult-onset seizures ... 37
Status epilepticus ... 37

## Chapter 4  Headache and Craniofacial Pain ... 38
Headache ... 38
Head pain ... 38
Migraine ... 38
Cluster headache ... 40
Tension headache ... 41
Sinus headache ... 41
Temporal arteritis ... 41
Trigeminal neuralgia ... 42
Post-lumbar puncture headache ... 42
Postconcussion syndrome ... 42
Pseudotumor cerebri ... 43
Headache from meningeal irritation ... 43

## Chapter 5  Toxic, Metabolic, and Nutritional Diseases ... 44
Alcohol poisoning ... 44
Methanol poisoning ... 49
Ethylene glycol poisoning ... 49
Opiate abuse ... 49
Psychostimulant toxicity ... 50
Lead poisoning ... 50
Arsenic poisoning ... 50
Mercury poisoning ... 50
Vitamin A excess ... 50
Vitamin E deficiency ... 51
Heat stroke ... 51
Hepatic encephalopathy ... 51
Uremia ... 51
Carbon monoxide poisoning ... 52
Hypoglycemic encephalopathy ... 52
Kernicterus ... 52
Acute intermittent porphyria ... 53

## Chapter 6  Dementia and Neurodegenerative Diseases ... 54
Dementia ... 54
Treatable dementia ... 54
Normal pressure hydrocephalus ... 54
Depressive pseudodementia ... 55
Vascular dementia ... 55
Multi-infarct dementia ... 55
Lacunar state ... 55
Neurodegenerative dementia ... 56
Alzheimer's disease ... 56

Pick's disease .................................................................................. 57
Infectious dementia ........................................................................ 57
Creutzfeldt-Jakob disease .............................................................. 57
AIDS dementia ............................................................................... 58
Progressive multifocal leukoencephalopathy .............................. 58
Spinocerebellar degeneration ....................................................... 59
Friedreich's ataxia ......................................................................... 59
Olivopontocerebellar atrophy ....................................................... 59
Ataxia-telangiectasia ..................................................................... 59

**Chapter 7 Coma and Impaired Consciousness** ............................... 61
Impaired consciousness ................................................................ 61
Terms related to coma and altered consciousness .................... 61
Glasgow Coma Scale ...................................................................... 62
Evaluation and treatment of coma or altered consciousness ... 63
Herniation syndromes ................................................................... 66
Brain death ..................................................................................... 68

**Chapter 8 Central Nervous System Trauma** ................................... 69
Head trauma ................................................................................... 69
Missile injuries ............................................................................... 69
Closed head injury ......................................................................... 70
Epidural hematoma ....................................................................... 71
Subdural hematoma ...................................................................... 71
Treatment of head injury .............................................................. 73
Delayed complications of head injury ......................................... 74
Spinal cord injury .......................................................................... 75

**Chapter 9 Peripheral Nervous System Disorders** ......................... 78
Normal peripheral nervous system ............................................. 78
Clinical terminology ...................................................................... 78
Clinical symptomatology .............................................................. 80
Signs ................................................................................................ 80
Electrodiagnostic tests .................................................................. 80
Pathophysiology of nerve disease ................................................ 81
Common polyneruopathies ........................................................... 82
Guillain-Barré syndrome ............................................................... 82
Motor neuropathies ........................................................................ 85
Amyotrophic lateral sclerosis ....................................................... 85
Charcot-Marie-Tooth disease ........................................................ 86
Poliomyelitis ................................................................................... 87
Focal neuropathies, plexopathies, and radiculopathies ............ 87
Carpal tunnel syndrome ............................................................... 87
Neuromuscular junction diseases ................................................ 93
Myasthenia gravis .......................................................................... 93
Muscle diseases .............................................................................. 96
Polymyositis .................................................................................... 96
Duchenne muscular dystrophy .................................................... 96

**Chapter 10 Dizziness and Vertigo** ..................................................... 100
Terminology .................................................................................... 100
Functional testing of vestibuloacoustic nerve ........................... 101
Vertiginous disorders .................................................................... 103
Ménière's disease ........................................................................... 104

Hyperventilation syndrome ... 106
Syncope and fainting ... 106

## Chapter 11  Central Nervous System Neoplasms ... 108
Pathophysiology of symptoms ... 108
Diagnosis of brain tumor ... 109
Glioma ... 110
Medulloblastoma ... 111
Retinoblastoma ... 111
Pineal region tumors ... 112
Pituitary adenoma ... 113
Colloid cyst of third ventricle ... 115
Meningioma ... 115
Schwannoma (neurilemoma) ... 115
Hemangioblastoma of cerebellum ... 116
Primary central nervous system lymphoma ... 116
Metastatic tumors ... 116

## Chapter 12  Movement Disorders ... 119
Diseases of the motor system ... 119
Features of abnormal motor activity ... 119
Parkinson's syndrome ... 121
Huntington's disease ... 123
Wilson's disease ... 123
Cerebellar ataxia ... 124
Gilles de la Tourette syndrome ... 124

## Chapter 13  Central Nervous System Infections ... 125
Central nervous system infections ... 125
Cerebrospinal fluid (CSF) examination ... 125
Bacterial meningitis ... 128
Syndrome of inappropriate ADH secretion ... 130
Tuberculous infection ... 133
Brain abscess ... 134
Fungal infection ... 135
Neurosyphilis ... 135
Lyme disease ... 137
Toxoplasmosis ... 137
Cysticercosis ... 138
Rickettsial infection ... 138
Aseptic meningoencephalitis ... 138
Herpes simplex encephalitis ... 139
Poliomyelitis ... 139
Rabies ... 139
Herpes zoster (shingles) ... 140
AIDS ... 140
Progressive multifocal leukoencephalopathy ... 141
Creutzfeldt-Jakob disease ... 141
Subacute sclerosing panencephalitis ... 142
Tetanus ... 142

## Chapter 14  Spinal Column Disease ... 144
Anatomic considerations ... 144
Spinal anomalies ... 144

| | |
|---|---:|
| Visceral diseases | 145 |
| Inflammatory spine disease | 145 |
| Spinal infection | 145 |
| Meningeal irritation | 146 |
| Intervertebral disc protusion | 147 |
| Spinal stenosis | 148 |
| Spinal tumor | 149 |
| Spinal cord infarction | 150 |
| Syringomyelia | 150 |
| Psychogenic pain | 151 |
| **Chapter 15  Sleep Disorders** | **152** |
| Sleep disorders | 152 |
| Normal sleep physiology | 152 |
| Biologic rhythms | 153 |
| Parasomnias | 154 |
| Insomnia | 156 |
| Narcolepsy | 157 |
| Sleep apnea | 158 |
| Circadian rhythm sleep disorders | 159 |
| **Chapter 16  Multiple Sclerosis and Demyelination** | **160** |
| Multiple sclerosis | 160 |
| Experimental allergic encephalomyelitis | 162 |
| Acute disseminated encephalomyelitis | 163 |
| Acute transverse myelitis | 164 |
| Central pontine myelinolysis | 164 |
| **Chapter 17  Developmental Disorders** | **166** |
| Developmental milestones and signs | 166 |
| Cerebral palsy | 167 |
| Congenital intrauterine infections | 168 |
| Mental retardation | 168 |
| Neurogenetic disorders | 169 |
| Hydrocephalus and spinal dysraphism | 170 |
| Anencephaly | 171 |
| Attention deficit hyperactivity disorder | 171 |
| Developmental specific learning disabilities | 171 |
| Selected additional reading | 172 |
| Self-assessment examinations | 174 |
| Answers to self-assessment examinations | 186 |

# Chapter 1    NEUROLOGIC EXAM AND NEURODIAGNOSTIC TESTS

I.  **History taking** — most important part of neurologic evaluation; provides information that permits examiner to arrive at tentative diagnosis (or diagnoses) which can then be verified or refuted by neurologic examination and neurodiagnostic testing

II. **Neurologic examination** — since nervous system **structure and function are related**, demonstration of functional abnormality provides information concerning localization of nervous system structures involved in disease process; certain examination basics include:

   A. **Mental status assessment**

   1. **Level of consciousness (alertness)**

      a. **Alert, awake, and oriented** — **normal** state of consciousness

      b. **Confusion (confusional state)** — **responsive** to stimuli, but **disoriented** with respect to time, person, and place

         (1) **Inattention** — inability to select among sensory stimuli, resulting in nearly random responses

         (2) **Delirium** — **acute** confusional state; sometimes also associated with drowsiness

      c. **Drowsiness** — **inclination to sleep; easily aroused** and able to respond to most stimuli both verbally and with motor defenses (fending off stimulus)

      d. **Stupor** — **little spontaneous physical or mental activity; reduced responsiveness** to environmental stimuli; generally **unresponsive to verbal stimuli**, and only **partially arousable** to vigorous (usually painful) stimulation; minimal or no verbal response, but able to respond with motor defenses

      e. **Semicoma (light coma)** — **appears to be sleeping**, but **generally unresponsive** to all but most vigorous stimulation; only **primitive reflexes** and abnormal body posturing occur

      f. **Coma (deep coma)** — lack of response to most painful stimuli; absence of primitive reflexes

Chapter 1

2. **Behavior** — **appropriateness** (or evidence of anxiety, depression, or personality disturbances)

3. **Memory**

   a. **Remote memory** — ability to recall events in distant past

   b. **Recent memory** — ability to relate recent events and to retain newly learned information

4. **Language** — **aphasia** (language disturbance) or **dysarthria** (abnormal speech articulation, but normal language ability); includes **spontaneous speech**, general **fluency**, **comprehension**, ability to **repeat** and **name** objects, **reading and writing** ability, and **handedness**

B. **Cranial nerve function**

   1. **Optic nerve (cranial nerve II)**

      a. **Visual acuity** — reduced acuity can be due to **refraction** problems, **obstruction** of light path to retina (such as occurs with cataracts), or **retinal** or **optic nerve** disease

      b. **Visual fields**

         (1) Bedside examination can identify gross visual field defects involving loss of half (**hemianopsia**) or quarter (**quadrantanopsia**) of visual field

         (2) Homonymous hemianopsia is loss of same half of visual field in each eye, while bitemporal hemianopsia is loss of temporal fields in each eye

         (3) Perimetry is necessary to detect small **scotoma** (small spots of visual loss) or changes in size of blind spot

Visual field defects: optic nerve [1]; near chiasm [2]; chiasm (bitemporal hemianopsia) [3]; optic tract (homonymous hemianopsia) [4]; temporal lobe optic radiations [5]; occipital lobe [6].

(4) Extinction or **neglect** — perception of individual stimulus in each half visual field, but perception of only one such stimulus when simultaneous bilateral stimuli are presented; suggests **parietal lobe lesion** contralateral to ignored half visual field

    c. **Funduscopy** — examination of optic nerve and retina with ophthalmoscope provides "window into the brain"

        (1) **Papilledema** — caused by **increased intracranial pressure** which impedes retinal venous return; earliest sign is **loss of venous pulsations**, followed by **blurring of optic disc margins, venous engorgement, elevation of optic disc**, and appearance of **retinal hemorrhages** (flame hemorrhages); visual acuity is usually spared

        (2) **Papillitis (optic disc inflammation)** — similar appearance to papilledema, except **visual acuity is lost**, and process is often unilateral (compared to papilledema which is usually bilateral)

        (3) **Optic atrophy** — **small pale sharply marginated optic disc**; end stage of various diseases affecting eye or optic nerve and usually associated with reduction in visual acuity

        (4) Normal variants

            (a) **Myelinated nerve fibers** — white area fanning out from optic disc along retinal nerve fibers, sometimes confused with "cotton wool spot" retinal infarcts

            (b) **Optic nerve drusen** — small white refractile bodies elevating optic nerve and blurring optic disc margins, sometimes confused with papilledema; termed **pseudopapilledema**

2. **Oculomotor nerve (cranial nerve III), trochlear nerve (cranial nerve IV), and abducens nerve (cranial nerve VI)**

    a. Eye movements controlled by six extraocular muscles

        (1) Superior rectus, medial rectus, inferior rectus, and inferior oblique are innervated by oculomotor nerve

        (2) Lateral rectus is innervated by abducens nerve

        (3) Superior oblique is innervated by trochlear nerve

# Chapter 1

    b. Eyelid elevation and pupillary size, shape, and reactivity to light and accommodation are controlled by oculomotor nerve

    c. **Nystagmus — abnormal involuntary rhythmic eye movement** at rest or induced by attempted voluntary eye movement; consists of **slow deviation** of eye **in one direction** and **quick corrective movement** in **opposite direction** (by convention, nystagmus is **named for quick component**)

        (1) **"End point" nystagmus** — occurs normally with gaze too far laterally

        (2) **Asymmetric lateral nystagmus** — nystagmus in only one direction of lateral gaze; indicates either vestibular or central nervous system abnormality

        (3) **Up-beating nystagmus, down-beating nystagmus,** or **rotatory nystagmus** — indicates brain stem disease

        (4) **Congenital nystagmus** — nystagmus present at birth that is horizontal-beating (even on upgaze) and is reduced or disappears with convergence; usually associated with reduced visual acuity

    d. **Oculomotor nerve (cranial nerve III) palsy**

        (1) **Complete oculomotor nerve (third nerve) palsy**

            (a) Eye is **abducted and deviated downward (down and out)**

            (b) Upper eyelid droops (**ptosis**)

            (c) **Pupil is dilated** and unreactive to light

        (2) **Partial oculomotor nerve (third nerve) palsy** — since autonomic pupillary constrictor fibers are peripherally arranged in nerve, partial damage can be differentiated by dissociation between eye movement and pupillary abnormalities

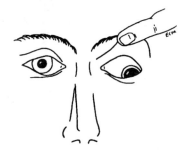

Complete third nerve palsy with eye deviated down and out and pupil dilated.

Internuclear ophthalmoplegia: adducting eye cannot go past midline, while abducting eye shows nystagmus; differentiated from medial rectus palsy by normal convergence.

(a) **Diabetic vascular disease (oculomotor nerve infarct** from occlusion of vasonervorum) **spares pupillary fibers**

(b) **Compression of oculomotor nerve** by transtentorial herniation of temporal lobe affects peripheral pupillary fibers first resulting in **pupillary dilation** before paralysis of extraocular muscles

e. **Abducens nerve (cranial nerve VI) palsy** — eye is **adducted** and does not move laterally beyond midline

f. **Internuclear ophthalmoplegia**

(1) Superficially resembles paralysis of medial rectus muscle except that ability to converge (look at near objects) is spared

(2) When attempting to look laterally, *ab*ducting eye moves laterally and develops **nystagmus** while *ad*ducting eye does not move past midline

(3) Results from **lesion of medial longitudinal fasciculus** in central part of pons

g. **Trochlear nerve (cranial nerve IV) palsy**

(1) Normal superior oblique muscle function keeps visual image upright by rotating eye slightly during minor sideward head movement

(2) Trochlear palsy results in **vertical diplopia (double vision)**, especially when looking downward at relatively near objects (as when climbing up stairs)

(3) Unilateral trochlear nerve palsy often results in **head tilt** (toward side opposite paretic superior oblique muscle) to prevent this diplopia

(4) Isolated trochlear nerve palsy is frequently due to head trauma

h. **Argyll Robertson pupil** — small, **irregular pupil** that **does not dilate** to mydriatic drugs or **constrict (react)** to light, but **constricts on accommodation** (characteristic of neurosyphilis: "accommodates, but doesn't react")

i. **Horner's syndrome**

(1) Frequently confused with oculomotor palsy, but actually due to **damage to sympathetic nerve supply** to eye

(2) Characterized by **small pupil (miosis), slight drooping of eyelid (ptosis)**, and **impaired sweating (anhydrosis)** on that side of face

(3) Damage can occur anywhere along course of sympathetic pathway from lateral medulla to globe; common sites of lesions include: **medullary infarction (Wallenberg syndrome)**, **T1 root injury (Klumpke-Dejerine palsy)**, **apical lung tumor** involving brachial plexus **(Pancoast's tumor)**, and **carotid bifurcation atherosclerosis** (damaging sympathetic nerve fibers in carotid artery sheath)

3. **Trigeminal nerve (cranial nerve V)** — provides corneal and facial sensation and innervates muscles of mastication; sensory divisions are ophthalmic (V1), maxillary (V2), and mandibular (V3)

    a. **Corneal reflex** — lightly touching cornea normally causes brisk **bilateral eye closure**; reflex is integrated in pons and motor component (lid closure) is mediated through facial nerve (cranial nerve VII) innervation of orbicularis oculi muscles

    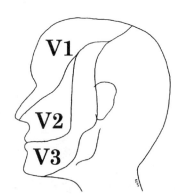

    Sensory distributions of three divisions of trigeminal nerve (cranial nerve V).

    (1) **Blink will not occur on side of facial nerve paralysis**, regardless of which cornea is touched, although patient will report that sensation is same in both corneas

    (2) Touching cornea on side of **ophthalmic division dysfunction will not result in eye blink**, but touching opposite cornea will cause bilateral eye blink; patient will report difference in corneal sensation

    b. **Facial sensation** — loss of sensation can occur independently in each trigeminal nerve division

    c. **Muscles of mastication** — atrophy of temporalis or masseter muscles can be evident on inspection; unilateral weakness during jaw opening causes **jaw deviation** *toward* **weak side**

4. **Facial nerve (cranial nerve VII)** — mediates **facial expression;** also supplies lacrimal gland (tearing), **stapedius muscle** (dysfunction results in sensitivity to sound or **hyperacusis**), salivary glands, and taste on anterior two-thirds of tongue

    a. **Lower motor neuron facial palsy** — lesion of **brain stem facial nucleus or facial nerve** produces weakness of **all facial muscles (including forehead)** on same side as lesion (unilateral complete facial weakness)

b. **Upper motor neuron facial palsy** — lesion **above** brain stem facial nucleus involving either **contralateral motor cortex or corticobulbar fibers**; unilateral **lower facial weakness** evident as **preserved ability to wrinkle forehead** (and partially close eyelids), but paralysis of remainder of facial muscles on one side (weakness in muscles of lower half of that side of face)

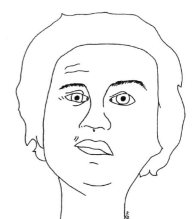

Left lower motor neuron facial palsy producing weakness of all facial muscles.

5. **Vestibuloacoustic nerve (auditory nerve; vestibulocochlear nerve; cranial nerve VIII)** — evaluation of **auditory acuity** and **vestibular function**

   a. **Weber's test** — with vibrating tuning fork placed in **midline on forehead**, sound should be heard equally in both ears; with unilateral damage to cochlea or nerve (**sensorineural deafness**), sound is heard better in ear with normal acuity, while with middle ear or outer ear disease (**conduction deafness**), sound is heard better in diseased ear

   b. **Rinne test** — vibrating tuning fork first held in front of external auditory meatus (**air conduction**) and then stem is placed firmly against mastoid process (**bone conduction**); bone conduction is louder than air conduction if patient has **conductive deafness** (middle or outer ear disease); air conduction is louder than bone conduction in normal individuals and those with **sensorineural deafness**

   c. **Nylen-Bárány maneuver** — patient seated on examining table is suddenly lowered to supine position with the head thrust 45 degrees backward over end of table and turned 45 degrees to one side; development of nystagmus (particularly asymmetric) and/or vertigo suggests vestibular disease

   d. **Caloric test** — **unilateral vestibular stimulation** is accomplished by instillation of cold or warm water into one external auditory meatus to induce **asymmetric nystagmus**; test is performed with patient lying flat and head flexed 30 degrees in order to stimulate **horizontal semicircular canal**

      (1) Warm water — slow movement of eyes away from stimulated ear with fast correction toward stimulated ear

      (2) Cold water — slow movement of eyes toward stimulated ear and fast correction away from stimulated ear

(3) Mnemonic for direction of nystagmus with caloric test — *COWS* (*c*old *o*pposite, *w*arm *s*ame)

(4) Effect of **cold water** is similar to **destructive lesion** of vestibular apparatus or vestibular nerve; effect of **warm water** is same as **irritative lesion**

6. **Glossopharyngeal nerve (cranial nerve IX)** and **vagus nerve (cranial nerve X)** — mediate pharyngeal reflexes (**gag reflex**), **phonation**, palatal sensation and movement, and **voice**; vagus nerve also supplies parasympathetic innervation and sensory fibers to thoracic and abdominal viscera

   a. Unilateral vagal nerve dysfunction results in palatal asymmetry with **uvula deviating toward normal side**

   b. Palatal weakness results in **nasal quality to voice**, while vocal cord weakness results in **hoarseness**

7. **Spinal accessory nerve (cranial nerve XI)** — innervation of sternocleidomastoid and trapezius muscles; **head turning to *right* is accomplished by *left* sternocleidomastoid** and turning to left is accomplished by right sternocleidomastoid

8. **Hypoglossal nerve (cranial nerve XII)**

   a. With unilateral hypoglossal nerve lesion, **tongue deviates toward weak side**; long-standing lesion results in atrophy of affected side of tongue

   b. **Motor neuron diseases** are associated with **fasciculations** (involuntary twitching of muscle fibers comprising tongue) and bilateral atrophy resulting in wrinkled appearance, particularly at lateral margin of tongue

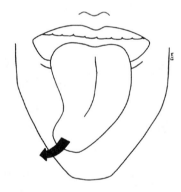

Tongue deviation *toward* weak side.

C. **Motor function**

1. **Gait and balance**

   a. Ability to rise from sitting position — abnormality occurs with proximal leg muscle weakness

   b. **Ordinary walking**

      (1) **Leg weakness** — suggests **muscle or peripheral nerve** disease

(2) **Ataxia (wide-based stance or unsteadiness)** — suggests **cerebellar disease, loss of proprioception** (vibratory and position sense), or **vestibular disease**

(3) **Stiffness** of legs with **circumduction** — suggests **spasticity**

(4) **Stooped posture** with small initial steps that become progressively faster (**festination**) — suggests parkinsonism

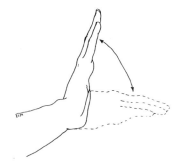

Sudden loss of tone in asterixis.

c. **Romberg test (falling** from standing position **after eye closure)** — suggests **loss of proprioception** (vibratory and position sense)

2. **Movements and postures**

   a. **Tremor** — Involuntary, **rhythmic oscillations** around joint

   b. **Myoclonus — Sudden, irregular, shock-like contractions** of one or more muscles

   c. **Asterixis** — frequent (several times per minute) arrhythmic **lapses of posture** associated with brief loss of electromyographic activity (electrical muscle silence)

   d. **Choreoathetosis** — chorea and athetosis are distinct movement disorders, but often occur together and individual movements blend into one another

      (1) **Chorea — purposeless, abrupt, rapid, irregular, involuntary, nonrepetitive, jerky movements**; to make movements less noticeable, patient attempts to incorporate choreic jerks into voluntary movement, but such superimposition makes voluntary movement appear exaggerated and awkward

      (2) **Athetosis — purposeless, slow, sinuous, writhing movements** which appear to flow into one another and interrupt attempts to maintain posture or to initiate voluntary activity

Spastic hemiparetic gait: stiff internally-rotated leg is thrust outward and forward (circumduction).

e. **Spastic hemiparetic posture — flexion of elbow and wrist**, adduction of shoulder, and extension and **internal rotation of leg** and feet; indicates **corticospinal tract damage**

f. **Ballismus** — purposeless, **violent, flinging** involuntary extremity movement, usually involving arm unilaterally (hemiballismus)

g. **Dystonia (torsion spasm) — sustained abnormal or inappropriate extreme posture**; body or limb positioning resembles an extreme form of athetosis in which changes of position are exceptionally slow

h. **Parkinsonian posture — forward flexion** of neck and body, **internal rotation of shoulders**, and arms hanging at side (palms facing backwards)

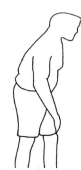

Parkinsonian posture with forward flexion of neck and body, internal rotation of shoulders, and arms hanging at side with palms facing backwards.

i. **Tardive dyskinesia — involuntary repetitive purposeless movements** (resembling choreoathetosis) usually restricted to head and neck, particularly **oral-buccal-lingual** musculature (tongue protrusion, chewing, lip smacking, grimacing); develops during continuous treatment with **dopamine-blocking neuroleptic drugs**

3. **Motor strength, tone, and bulk**

   a. **Strength** is often graded according to British Medical Research Council **(MRC) scale**

   b. **Muscle tone** — evaluated by passively moving limbs

   | MRC SCALE OF MUSCLE STRENGTH | |
   |---|---|
   | 5 | Normal |
   | 4 | Apparent weakness |
   | 3 | Movement possible against gravity |
   | 2 | Movement possible only when force of gravity eliminated |
   | 1 | Flicker of muscle contraction (no joint movement) |
   | 0 | No muscle contraction |

   (1) **Rigidity — increased muscle tone** present **throughout range of motion** of limb, with **constant and uniform resistance to passive movement** (like that noted in attempting to bend a lead pipe); observed in **parkinsonism** and **diseases of basal ganglia**

   (2) **Spasticity — increased muscle tone** and increased tendon reflexes associated with damage to **pyramidal motor system (corticospinal tract)**; during rapid passive limb movement, there is an abrupt "catch" and **increasing resistance**

to movement, followed by abrupt reduction in resistance as movement continues ("clasp knife" phenomenon)

      (3) **Hypotonia** — **floppy** extremity; usually seen with severe peripheral nerve (denervating) disorders or cerebellar disease

  c.  **Atrophy** is found with **disuse, denervation,** or **muscle destruction** and usually correlates with degree of weakness; muscle **hypertrophy** is most commonly found with **Duchenne muscular dystrophy**

  d.  **Fasciculations**

      (1) **Brief visible muscle twitches** indicative of excessively **irritable motor nerve**, as occurs in **denervating diseases**

      (2) **Benign fasciculations** — presence of fasciculations in individuals without denervating disease; most often occurs following **strenuous muscle activity**, but also associated with **hyperthyroidism, excessive caffeine intake,** or administration of **acetylcholinesterase inhibitors** such as pyridostigmine (Mestinon)

4.  **Reflexes**

  a.  **Plantar reflex (response)**

      (1) **Stroking sole (plantar surface) of foot** with sharp object normally results in **flexion of toes**

      (2) **Babinski reflex** — **pathologic** response to **plantar stimulation** consisting of **extension (dorsiflexion) of big toe** and **fanning and extension (dorsiflexion) of other toes**; indicative of **pyramidal motor system (corticospinal tract)** lesion

Babinski reflex: plantar stimulation results in extension of big toe and fanning and extension of other toes.

  b.  **Tendon (muscle stretch) reflexes** — brief muscle contraction in response to sudden stretch produced by sharp blow to tendon

      (1) **Biceps reflex** — C5-C6 roots; musculocutaneous nerve

      (2) **Brachioradialis reflex** — C5-C6 roots; radial nerve

(3) **Triceps reflex** — C7-C8 roots; radial nerve

(4) **Knee (patellar) reflex** — L3-L4 roots; femoral nerve

(5) **Ankle (achilles) reflex** — S1 root; tibial nerve

(6) **Hyperactive reflex** — indicates lesion of **corticospinal tract (upper motor neuron** or **pyramidal motor system)** above segmental level of reflex; often associated with **clonus (rhythmic muscle contractions** precipitated by sudden muscle stretch) which is most readily observed at ankle

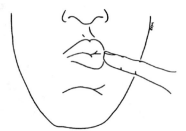

Lips turn toward stroking stimulus at corner of mouth in rooting reflex.

(7) **Hypoactive or absent reflex** — indicates lesion involving some part of **reflex arc**: including sensory endings of muscle spindle, sensory nerve, spinal cord dorsal and ventral (anterior) horns, motor nerve, neuromuscular junction, or muscle fibers; reduction or loss of reflexes occurs early in sensory nerve disease and later in motor nerve or muscle fiber disease

(8) **Reflex asymmetry** suggests unilateral lesion; rostral-caudal gradient (hyperactive in legs compared to arms) suggests spinal cord lesion

c. **Release signs**

(1) **Rooting reflex** — lips turn toward stroking stimulus at corner of mouth

(2) **Grasp reflex** — grasp elicited by lightly stroking palm

(3) **Snout reflex** — puckering of lips in response to gentle tapping of upper lip

5. **Sensation**

a. **Pinprick (pain) sensation** and **proprioception** (vibratory and joint position sense)

b. **Cortical sensation** — ability to identify objects in hand (**stereognosis**) and numbers written on hand (**graphesthesia**)

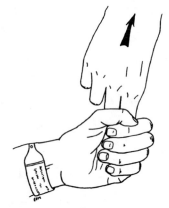

Lightly stroking palm causes patient to grasp.

# NEUROLOGIC EXAM AND NEURODIAGNOSTIC TESTS

III. **Lumbar Puncture**

   A. Procedure (performed by attending physician) for obtaining **cerebrospinal fluid** which can be used for **biochemical analysis, cellular examination,** and **culture**

   B. Performed to evaluate:

   1. **Inflammation** and **organisms** in suspected central nervous system infection (**meningitis or encephalitis**)

   2. Red blood cells in **subarachnoid hemorrhage**

   3. Protein abnormalities in **Guillain-Barré syndrome (albuminocytologic dissociation)** and **multiple sclerosis (oligoclonal bands and elevated myelin basic protein and γ-globulin)**

   4. Cytologic abnormalities in **leukemia** and **lymphoma**

   5. **Serologic evidence** of infection in neurosyphilis

   6. Pressure measurement in **pseudotumor cerebri**

   7. To **introduce drugs** into subarachnoid space for treatment of cancer or to **introduce contrast agents** in evaluation of mass lesions

   C. Site of puncture is usually at **L3-L4 or L4-L5 interspinous space**; spinal needle is inserted using **sterile technique**

   D. Normal values:

   1. **Opening pressure** should be **no more than 180 mm of water**

   2. **Thin colorless sparkling crystal-clear fluid** that does not coagulate

   3. **Glucose** level should be greater than 40 mg/dL and about **60% of blood glucose level**

   4. **Protein** level should not be over 45 mg/dL in young adults

   5. **Cell count** should reveal *no neutrophils* or other polymorphonuclear leukocytes and *no more than five lymphocytes* per cubic millimeter (microliter)

   6. Fluid should be sterile when cultured

## Chapter 1

E. Possible complications include **post-lumbar puncture headache** from persistent fluid leakage through hole in arachnoid and dura and **iatrogenic meningitis** from inadequate sterile precautions

IV. **Computed tomographic (CT) scans**

A. **Computer-generated image** of brain or spine; based upon detection of degree of attenuation (absorption) of collimated x-ray beams passing through body tissue; images are slices through brain or body, usually in axial plane (perpendicular to vertical axis of body); identification of normal structures and anatomical localization of lesions possible with this technique; particularly valuable for demonstrating dense (calcified) lesions or processes that destroy tissue (such as necrosis) and thus decrease tissue density; intravenous infusion of iodinated contrast solutions increases density of blood vessels, vascular tumors, and areas in which blood-brain barrier has been disrupted

B. Useful in evaluating focal neurologic deficits, altered mental status, head trauma, increased intracranial pressure, and suspected mass lesions

V. **Magnetic resonance imaging (MRI)**

A. **Computer-generated image** of brain or spine providing much greater detail than CT scan; performed by placing patient in powerful magnetic field which causes hydrogen atoms (protons) to reorient; subsequent activation by radiofrequency pulse causes protons to resonate (relaxation), emitting signals which are computer-analyzed to generate images in sagittal, coronal, and horizontal planes

B. Variations in type of activating radiofrequency signals and in recording of subsequently emitted signals can produce different images; typical images are $T_1$-weighted, $T_2$-weighted, or intermediate (proton-density or spin density); major characteristics of different MRI images include:

1. Bone or areas of calcification are dark (this is in contrast to CT scans which show bone or calcification as white areas of signal attenuation)

2. Cerebrospinal fluid (or water) is dark in $T_1$-weighted images and bright (white) in $T_2$-weighted images

C. Intravenous infusion of gadolinium (paramagnetic contrast agent) increases signal (in $T_1$-weighted images) from highly vascular tissues or areas of blood-brain barrier breakdown

D. Best technique for visualizing brain and spinal cord anatomy and structural lesions; however, CT scans are best for evaluation of bony anatomy

E. **Magnetic resonance angiography (MRA)** — specialized computer processing of signals permits visualization of blood vessels that mimics images of cerebral angiography

VI. **Myelography**

　A. Instillation of **iodinated contrast into spinal subarachnoid** space permits visualization of anatomical relationships of spinal cord to surrounding structures in spine radiographs or spinal CT scans

　B. Particularly useful in identifying mass lesions causing spinal cord or root compression

VII. **Cerebral angiography (arteriography)**

　A. Through catheter inserted into femoral artery and threaded up to carotid or vertebral arteries, **iodinated contrast material is infused** while simultaneous **rapid sequence radiographs** of neck and head are obtained; this permits detailed visualization of brain vascular supply

　B. Particularly valuable for study of aneurysm, arteriovenous malformation, transient ischemic attack (TIA), cranial arteritis, or vascular tumor prior to surgery

VIII. **Isotope cisternography** — injection of **radioactive material into lumbar subarachnoid space** in order to follow its circulation and elimination from cerebrospinal fluid space; particularly useful in study of normal pressure hydrocephalus (NPH) and cerebrospinal fluid leaks (otorrhea or rhinorrhea)

IX. **Electroencephalography**

　A. **Multiple scalp electrodes** recording of spontaneous electrical activity produced by cerebral neurons; activity normally varies in different regions, in different states of arousal and sleep, and with age

　B. Abnormal activity is recorded as **changes in amplitude (voltage) or frequency** of waves; **spikes and sharp waves are epileptiform** (characteristic of seizures); **slow waves are found with metabolic disturbances or destructive processes**

　C. Particularly useful in evaluating epilepsy, coma, dementia, sleep disorders, and brain death

X. **Electromyography (EMG)**

　A. Recording of **electrical activity of muscle fibers** through needle electrode inserted into muscle belly; changes in potentials generated by single motor units, as well as abnormal spontaneous electrical activity (such as **fasciculations** or **fibrillations**) can be detected

Chapter 1

B. Particularly useful in identifying neurogenic changes (denervation or reinnervation) in muscles; localizes site of lesion by delineating distribution of changes in different muscles; recovery from nerve injury can be predicted or confirmed by serial studies

XI. **Nerve conduction velocity (NCV) study**

A. **Stimulation of motor or sensory nerve** at different points and **calculation of velocity** of conduction of propagated impulse; measures **velocity of fastest conducting fibers (large myelinated axons); normal velocity if myelin is intact**; either demyelination (with preservation of axons) or destruction of large myelinated fibers can result in slow conduction velocity; **F-wave latency** evaluates conduction velocity in **proximal** nerve and nerve root; **H-reflex** evaluates **reflex arc** in lower extremity

B. Particularly useful in studying demyelinating polyneuropathy (Guillain-Barré syndrome) or focal demyelination (as occurs in entrapment neuropathies such as carpal tunnel syndrome)

XII. **Audiometry**

A. Hearing tests include **pure tone audiometry, speech discrimination, alternate binaural loudness balance (ABLB), short increment sensitivity index (SISI), tone decay, impedance tympanometry, stapedial reflex**, and **Békésy audiogram**

B. Particularly useful in defining auditory function in patients with complaints of dizziness or vertigo

XIII. **Evoked Potentials**

A. Skin (surface) electrode recording and computer analysis of central nervous system electrical activity following **repeated stimulation** of peripheral sensory receptors

1. **Brain stem auditory evoked response (BAER)** — stimulus of clicks delivered through earphones to define potentials in auditory pathway

2. **Visual evoked potentials** — stimulus of stroboscopic flash or changing checkerboard pattern on television screen (pattern shift visual-evoked potentials) to define potentials in visual pathway

3. **Somatosensory evoked potentials** — small amplitude electric stimuli delivered to peripheral nerves to define potentials in somatosensory pathways including spinal cord and brainstem

B. Particularly useful in defining diseases involving white matter tracts (such as multiple sclerosis); also used to monitor brain stem or spinal cord integrity in anesthetized patients during neurosurgical procedures and to evaluate intactness of brain stem in coma

# NEUROLOGIC EXAM AND NEURODIAGNOSTIC TESTS

XIV. **Neurosonography (ultrasonography)**

    A. Computer images produced by analysis of echoes of **high frequency sound** (generated by device placed on skin surface) deflected from interfaces between structures of different density; duplex scans couple Doppler signals to investigate blood flow in vessels

    B. Particularly useful in evaluating prenatal, neonatal, and infant brain abnormalities since poorly mineralized skull or open fontanelles permit imaging (sound wave dispersion by mineralized cranial bones makes ultrasonography useless once fontanelles close); duplex scans provide non-invasive information concerning vascular patency (such as degree of carotid artery stenosis)

XV. **Muscle biopsy** — histochemical study of **snap-frozen muscle biopsy specimens** can differentiate neurogenic from myopathic disorders and characterize specific myopathies; **biochemical analysis** of muscle specimens can define genetic abnormalities in enzymes or proteins (such as dystrophin deficiency in Duchenne muscular dystrophy)

# Chapter 2 — STROKE

I. **Terminology**

  A. **Stroke** — **rapidly developing focal neurologic deficit persisting more than 24 hours**; implies **vascular etiology** (either hemorrhagic or ischemic) with **brain tissue destruction**

  B. **Transient ischemic attack (TIA)** — **rapidly developing focal neurologic deficit resulting from ischemia** but persisting for **less than 24 hours**; implies only brain tissue **dysfunction** with **no tissue destruction**

  C. Other causes of focal neurologic deficit must be excluded, such as **subdural hematoma**, **abscess**, **tumor**, or **epilepsy**; vascular nature of events must be proven and etiology determined (such as atherosclerotic occlusion, thromboembolism, vasculitis, ruptured aneurysm, or arteriovenous malformation)

  D. **Ischemic stroke** — **interruption of blood flow** to area of brain tissue resulting in tissue death

  E. **Infarction** — **tissue destruction (necrosis)** secondary to lack of blood supply

    1. **Pale (bland, white) infarction** — tissue necrosis (**devoid of blood**)

    2. **Hemorrhagic (red) infarction** — extensive **leakage of red blood cells** from vessels in and around necrotic tissue (implies cessation of blood flow long enough to produce death of neural tissue and damage to vascular endothelium, with subsequent restoration of blood flow and consequent extravasation)

  F. **Hemorrhagic stroke** — **hemorrhage** with consequent destruction of brain tissue

II. **Lesion localization** — nervous system structure and function are interrelated: functional abnormalities (identified neurologic deficits) indicate area of brain involvement; structural changes identified on imaging indicate expected neurologic deficits

  A. **Cerebral hemispheres** — **paralysis** and **sensory disturbance** involving contralateral face, arm, and leg; **aphasia** (language disturbances) if left hemisphere involved; initially only **minimal reduction in level of consciousness** (although with increasing edema and herniation, consciousness may be altered)

  B. **Posterior fossa (brain stem or cerebellum)** — frequently associated with **crossed deficits** (ipsilateral abnormality in face and contralateral involvement of rest of body); **speech is slurred** or unintelligible (**dysarthric**), but language is normal (**no aphasia**); **early alteration in consciousness** more common than with cerebral hemisphere lesions

III. **Clues in diagnosis**

    A. **Thromboembolic stroke** — **sudden onset** with no warning and with **maximal deficit from onset**; most often occurs during waking hours; deficit frequently improves during first few days

    B. **Hemorrhagic stroke** — **gradually developing** (over minutes to hours) deficit often associated with severe **headache, nausea, vomiting** and sometimes with **stiff neck**; usually occurs during waking hours; often individual is known to have **hypertension** or **bleeding tendency**

    C. **Thrombotic stroke** — frequently **preceded by warning** such as prior **transient ischemic attacks (TIAs)**; neurologic deficit usually **progresses in stepwise fashion** (sudden worsening, stability, further sudden worsening, etc.) over several hours or days ("**stroke in evolution**"); onset often occurs overnight during sleep with deficit evident upon awakening in morning

        1. **Carotid distribution TIA** — common symptoms include transient **visual loss** in one eye (**amaurosis fugax**), unilateral weakness or sensory disturbances, or aphasia

        2. **Vertebrobasilar distribution TIA** — common symptoms include **vertigo** or dizziness, **ataxia**, dysphagia, **sudden loss of consciousness, diplopia**, or sensory disturbances in face

    D. **Lacunar stroke**

        1. Associated with **hypertension** in which **hyalinization (lipohyalinosis) of small penetrating arteries** results in small **cystic infarcts** affecting (in order of frequency) **putamen, caudate, thalamus, basis pontis**, and **internal capsule**

        2. Common lacunar stroke syndromes include:

            a. **Pure motor hemiplegia** — weakness of face, arm, and leg (without any other symptoms); due to lacunar strokes in **posterior limb of internal capsule** or **basis pontis**

            b. **Pure sensory stroke** — paresthesias of limbs and face, but only minimal objective sensory loss and no other symptoms; due to lacunar strokes in **thalamus**

            c. **Dysarthria-clumsy hand syndrome** — facial weakness and severe dysarthria accompanied by hand clumsiness; due to lacunar strokes in **pons**

Chapter 2

   d. **Lacunar state**

   (1) **End-stage of numerous lacunar strokes** (usually following long-standing, poorly-controlled **hypertension**) resulting in **bilateral pseudobulbar palsy**

   (2) Characterized by:

   (a) Weakness of voluntary bulbar muscle activity resulting in **drooling, dysphagia, dysarthria**, and **emotional incontinence (uncontrolled reflex laughter and crying)**

   (b) Spastic **shuffling gait** and **bilateral Babinski reflexes**

E. **Subarachnoid hemorrhage** — sudden onset of **severe headache**, photophobia, and **stiff neck**; frequently occurs during physical exertion

F. **Retinal emboli** — cholesterol emboli in retinal vessels suggest disease of **carotid artery** in which **complicated atherosclerotic plaques** shower emboli

G. **Collateral flow** — slowly developing internal carotid artery occlusion is suggested by **increased pulses in ipsilateral external carotid artery branches** (frontal artery branch of superficial temporal artery and dorsal nasal artery branch of external maxillary artery) anastomosing with ophthalmic artery branches

H. **Horner's syndrome** — often accompanies **severe ipsilateral carotid artery atherosclerosis** due to damage to sympathetic nerve fibers traveling in carotid artery sheath

I. **Carotid artery bruit** — turbulence of blood flow in **severely stenotic carotid artery** is detectable by auscultation

J. **Central retinal artery pressure** — ophthalmodynamometry reveals **reduced pressure** when **carotid artery stenosis** is hemodynamically significant

IV. **Selected stroke syndromes**

A. **Middle cerebral artery stroke syndrome** — usually results from **thromboembolism to origin of middle cerebral artery** or **occlusion of internal carotid artery**; characterized by **hemiplegia** or hemiparesis (greatest in arm), **aphasia** (involving **left cerebral hemisphere**) or non-dominant parietal lobe syndrome (involving right cerebral hemisphere), unilateral cortical sensory loss **(hemianesthesia), homonymous hemianopsia**, and conjugate eye deviation (toward side of hemisphere lesion)

B. **Anterior cerebral artery stroke syndrome** — unilateral **leg weakness** and **cortical sensory loss, incontinence, release signs** (grasp, rooting, and snout reflexes), and **abulia** (lack of will power or initiative; extreme apathy)

C. **Posterior cerebral artery stroke syndrome** — **homonymous hemianopsia** (usually with **macular sparing**) can be only deficit (patient is often unaware of this deficit); other symptoms can include loss of recent memory, alexia without agraphia (inability to read, but preserved ability to write), or hemianesthesia

D. **Vertebrobasilar artery stroke syndrome** — numerous eponymic syndromes due to infarctions in territories of various small and large branches; selected syndromes include:

Lateral cerebral hemisphere showing relationship of motor homunculus to anterior cerebral (dark), posterior cerebral (gray), and middle cerebral (white) arterial territories.

   1. **Wallenberg syndrome (lateral medullary syndrome)** — unilateral occlusion of vertebral artery or posterior inferior cerebellar artery produces **infarction in dorsolateral medulla**; characteristic features include:

      a. Acute onset of severe **vertigo, nausea, vomiting, nystagmus, and oscillopsia**

      b. Signs of **gait** and *ipsilateral* **limb ataxia**

      c. **Loss of pain and temperature sense on** *ipsilateral* **face** (decreased corneal reflex) and *contralateral* **body**

      d. *Ipsilateral* **weakness of palate and vocal cords** with diminished gag reflex, **dysphagia**, and hoarseness

      e. *Ipsilateral* **Horner's syndrome** (ptosis, miosis, and facial anhidrosis)

      f. **Intractable hiccups** can also occur

      g. **No limb weakness or paralysis**

   2. **Basilar artery occlusion** — severity of signs varies depending on degree of collateral flow; characterized by bilateral corticospinal tract damage (**quadriplegia**) and sensory loss, along with various **cranial nerve palsies** and either **coma** or **locked-in syndrome**

Chapter 2

E. **Cerebellar stroke (infarction)**

1. Associated with severe **dizziness, ataxia, nystagmus,** and nausea and vomiting

2. Swelling from edema can compress brain stem resulting in **cranial nerve palsies**, corticospinal tract signs (**quadriplegia**), and **coma**

3. **Neurosurgical decompression** is often necessary to prevent **cerebellar tonsilar herniation** and death; survivors frequently have only minimal residual signs of ataxia

V. **Hypertensive hemorrhagic stroke (intraparenchymal hemorrhage; hypertensive intracerebral hemorrhage)**

A. Damage to **small penetrating arteries** from **chronic hypertension** results in **vascular hyalinization** and development of **small false aneurysms (Charcot-Bouchard aneurysms)** which presumably rupture producing intraparenchymal hemorrhage

B. Common sites of hemorrhage (in order of frequency) include: putamen, thalamus, cerebellum, and pons

1. **Putamen** — findings suggestive of middle cerebral artery stroke syndrome, but with **greater reduction in level of consciousness** (in part due to early transtentorial uncal herniation from mass effect of hemorrhage)

2. **Thalamus — hemiplegia and hemianesthesia with abnormal eye movements** due to extension of hemorrhage into or pressure on midbrain; abnormal eye movements include gaze palsies (inability to move eyes in specific direction of gaze), forced deviation of eye in particular directions (often downward), nystagmus, or skew deviation (vertical strabismus; one eye higher than other eye); early coma from central rostral-caudal herniation

3. **Cerebellum** — alert individual with sudden onset of **dizziness, headache, vomiting, inability to walk due to severe ataxia, and marked hypertension** are cardinal early features; nystagmus and eye movement disturbances can be evident on examination, but muscle strength and sensation are normal; if untreated, compression of brain stem and cerebellar tonsillar herniation rapidly results in coma and death; considered to be **medical emergency** necessitating prompt neurosurgical decompression, which can be life-saving; survivors often have only minimal residual ataxia

4. **Pons** — onset of **coma, decerebrate rigidity, pinpoint pupils (1 mm) that react to light, and absent vestibulo-ocular reflexes**; death usually occurs within several hours

C. Differentiation from ischemic strokes is possible with radiologic imaging studies

VI. **Subarachnoid hemorrhage**

  A. **Saccular (berry) aneurysm**

  1. Aneurysms develop at **arterial bifurcations** due to defect in vascular media and elastica

      a. Current theory of pathogenesis: at sites of congenital absence of vascular media, destruction of internal elastic membrane by atherosclerosis permits bulging and consequent aneurysm formation

      b. More common in patients with **coarctation of aorta** and **polycystic kidney disease**

      c. **Rarely evident before age 20 years**; most ruptures occur between **ages 35 and 65 years**; more common in **females**; may occur in up to 2% of population; affected individuals frequently (20%) have **more than one aneurysm**

  2. Most commonly located in anterior part of circle of Willis (in order of frequency): **anterior communicating artery, junction of posterior communicating and internal carotid arteries, first bifurcation of middle cerebral artery, and branching of internal carotid into middle and anterior cerebral arteries**

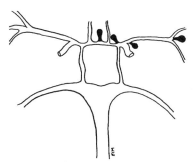

Common arterial aneurysm sites: anterior communicating, junction of posterior communicating and internal carotid, first bifurcation of middle cerebral, and trifurcation of internal carotid.

  3. Presents as **sudden violent headache** ("worst ever"), photophobia, and **stiff neck**; onset often associated with physical activity; loss of consciousness, confusion, or coma are infrequent and localizing signs are uncommon:

      a. Focal neurologic deficits can occur from intraparenchymal dissection of blood or from **vasospasm** (vascular constriction reducing blood flow, presumably related to presence of blood in subarachnoid space)

      b. Early **complete oculomotor nerve (cranial nerve III) palsy** suggests nerve compression by **posterior communicating artery aneurysm**

  4. **Recurrent (often fatal) hemorrhage** is common (rebleeding occurs within two weeks in 20%); thus, surgical treatment is indicated to prevent rebleeding; usual surgical approach is to clip neck of aneurysm

Chapter 2

        5.    **Hydrocephalus** can develop due to blockage of normal cerebrospinal fluid circulation from fibrosis in subarachnoid space following resolution of hemorrhage

  B.  **Arteriovenous malformations**

        1.    **Developmental abnormality** consisting of **tangle of dilated blood vessels** forming abnormal communication between arterial and venous systems; many **vessels are thin-walled** and prone to rupture

        2.    Presentations vary: **seizures, recurrent headaches** (mimicking migraine), or **hemorrhage** (parenchymal hemorrhage or subarachnoid hemorrhage)

        3.    Diagnosis made with radiologic imaging studies

        4.    Treatment:

            a.    Embolization with thrombogenic chemicals and epoxy plastics injected by arterial catheters under fluoroscopic control

            b.    Neurosurgical ligation of feeding vessels and resection of involved brain tissue

            c.    Radiation therapy to produce vascular thrombosis

  C.  **Mycotic aneurysms** — aneurysmal dilation of **arterial wall damaged by bacterial or fungal inflammation**; initiated by septic embolus (usually of cardiac or pulmonary origin) containing infectious organisms; typically involves multiple peripheral branches; treatment of underlying disorder (such as bacterial endocarditis) is necessary

VII.  **Inflammatory diseases of cranial arteries**

  A.  **Temporal arteritis (cranial arteritis; giant-cell arteritis)**

        1.    **Inflammatory disorder** of cranial arteries in **older individuals** (over age 50 years) presenting with headache, **thickened tender non-pulsatile temporal or other scalp arteries**, and **transient visual disturbances** (blurring or blindness) that progress to **permanent blindness**; patients also have symptoms of **polymyalgia rheumatica** (**malaise, weight-loss**, and generalized **muscle aches** and weakness)

        2.    **Granulomatous (giant-cell) inflammation** and occlusion involving extracranial arteries (particularly ophthalmic artery resulting in blindness); involvement of carotid arteries (and other major aortic branches) can also result in occlusion and consequent strokes; diagnosis established by histologic study of **temporal artery biopsy**

3. Characteristically associated with **marked elevation of erythrocyte sedimentation rate (ESR)**, often up to 120 mm/hour ("ESR greater than patient's age")

4. **Immediate high-dose corticosteroid** treatment is necessary to prevent permanent visual loss

B. **Meningovascular syphilis** — **strokes** due to **arteritis** produced by **neurosyphilis**; usually affects younger individuals; occurs 5-10 years after initial infection by syphilis organism

VIII. **Sickle-cell anemia** — inherited disorder due to **abnormal hemoglobin (hemoglobin S)** which causes red blood cells to deform (sickle) under conditions of low oxygen tension; **sickled cells sludge in small vessels** causing infarction in brain, bone, and visceral organs

IX. **Hypoxia**

A. **Reduced oxygen delivery to brain**; usually implies global cessation of brain perfusion (as in cardiac arrest); also, termed **hypoxia-ischemia**

B. Brain damage occurs after 5 minutes of severe hypoxia (anoxia), but hypothermia prolongs this period

C. **Selective vulnerability**

1. Susceptibility of cell types to damage (in order of greatest susceptibility): **neurons, oligodendrocytes, astrocytes,** and **capillary endothelial cells**

2. Certain **neuronal populations** are more sensitive than others: **cerebral cortical layers III, V, and VI, Sommer's sector of hippocampus,** amygdala, and **cerebellar Purkinje cells**

D. Survival following restoration of blood flow and oxygenation often results in residual neurologic signs; some specific patterns include:

1. **Laminar cortical necrosis** — **widespread destruction of vulnerable cerebral cortical layers**, visible as layer of cortical destruction in radiologic imaging studies; associated with **persistent coma**

2. **Destruction of hippocampus** (and amygdala) — associated with **dementia and memory impairment**

3. **Destruction of cerebellar Purkinje cells** — associated with **intention (action) myoclonus**

X. **Border zone (watershed) infarction** — occurs at border zones of collateral blood flow between territories of major cerebral arteries; most often due to **hypotension** in which blood flow is most

# Chapter 2

severely compromised in terminal branches (referred to as "law of the most distant field" in analogy to pattern of agricultural irrigation failure)

XI. **Aphasia and language disturbances**

   A. Disturbance of *language* only; aphasia must be differentiated from disorders of *speech* (dysarthria)

   B. **Broca's (non-fluent, motor, anterior, expressive) aphasia**

   1. **Non-fluent, effortful, ungrammatic, spoken language** composed of short **telegraphic poorly-articulated** phrases; abnormal speech melody and rhythm

   2. **Auditory comprehension is normal**; reading silently for meaning is preserved, but reading aloud is impaired; **poor repetition** of spoken words; writing is large, messy, and effortful, paralleling verbal expression; patient is often aware of and frustrated by expressive difficulties

   3. Results from infarction involving **left hemisphere inferior frontal gyrus (Broca's area)** extending deep to lateral ventricle and back to precentral gyrus; usually associated with **right hemiparesis**

   | Fluency | Comprehension | Repetition | APHASIA |
   |---|---|---|---|
   | + | + | + | Anomic |
   | + | + | − | Conduction |
   | + | − | + | Transcortical sensory |
   | + | − | − | Wernicke's |
   | − | + | + | Transcortical motor |
   | − | + | − | Broca's |
   | − | − | + | Mixed transcortical |
   | − | − | − | Global |

   C. **Wernicke's (sensory, posterior, receptive) aphasia**

   1. **Voluminous (fluent) verbal output** with normal speech rhythm and melody, but content is **incomprehensible (nonsensical)**

   2. **Auditory comprehension is poor**, but patient often can respond to whole-body commands (such as "stand up"); **poor repetition** of spoken words; reading aloud and reading silently for meaning are also defective; writing is well-formed, but contains same errors and lack of meaning as verbal output; patient generally is unaware of language problem

   3. Results from infarction involving Wernicke's area or **left posterior-superior temporal gyri and temporal operculum** (portion of temporal lobe, including planum temporale, buried in Sylvian fissure); usually accompanied by **right homonymous superior quadrantanopsia** (from involvement of Meyer's loop portion of optic radiations)

D. **Global aphasia**

   1. **Alert, awake individual** who is **silent** except for infrequent stereotyped responses (such as "I don't know")

   2. **Combination of Broca's aphasia and Wernicke's aphasia**; produced by extensive damage to area surrounding left Sylvian fissure (**perisylvian lesion**) which represents major portion of middle cerebral artery blood supply

# Chapter 3 — EPILEPSY AND SEIZURES

I. **Terminology** — seizure versus epilepsy

    A. Definitions

        1. **Seizure** — sudden alteration in brain function due to **abnormal, excessive electrical discharges** by cerebral neurons; symptom of disease, not disease itself

        2. **Epilepsy** — symptom complex (**syndrome or disease**) characterized by **tendency to have repeated seizures**; also termed **seizure disorder**

    B. **Everyone who has epilepsy has seizures, but not everyone who has seizures has epilepsy**

    C. **Neither seizures nor epilepsy can be considered final diagnoses**:

        1. **Seizure is symptomatic** of many different possible underlying conditions (only one of which is epilepsy)

        2. **Epilepsy symptom complex has many possible etiologies** ranging from idiopathic conditions to various identifiable brain disorders

---

**SEIZURE CLASSIFICATION**
- Generalized seizures
  - Primary
    - Absence seizure
    - Clonic convulsion
    - Tonic convulsion
    - Tonic-clonic convulsion
    - Atonic seizure
    - Myoclonic seizure
  - Secondary
    - Partial with secondary generalization
- Partial (localized, focal) seizures
  - Simple
    - Motor or sensory
  - Complex
    - Behavioral, psychomotor, temporal lobe

---

II. Characteristics of seizures

    A. **Generalized seizure** — seizure beginning bilaterally (both cerebral hemispheres)

    B. **Partial (localized, focal) seizure** — seizure beginning in part of one cerebral hemisphere; may progress to involve both hemispheres (**secondarily generalized seizure**)

        1. **Simple partial seizure** — consciousness not impaired

        2. **Complex partial seizure** — consciousness impaired

---

**EPILEPSY CLASSIFICATION**
- **Primary (idiopathic) generalized epilepsy**
  - Petit mal epilepsy
  - Grand mal epilepsy
- **Primary (idiopathic) partial epilepsy**
  - Benign centrotemporal epilepsy
- **Secondary generalized epilepsy**
  - Infantile spasms (West) syndrome
  - Lennox-Gastaut syndrome
- **Secondary partial epilepsy**
  - Simple partial epilepsy (focal motor, jacksonian)
  - Complex partial epilepsy (temporal lobe)

C. **Ictus** (ictal event) — seizure event

D. **Postictal** — period immediately following end of seizure (blends into interictal period); usually refers to reversible phenomena which follow seizure

E. **Interictal** — period **between seizures**

F. **Aura** — sensation (**warning**) that seems to precede clinical seizure; actually is beginning of ictal event

G. **Status epilepticus** — **continual seizures** lasting more than 30 minutes or (more commonly) **repeated seizures** in which individual does not recover between seizures

| SELECTED DISEASES WITH EPILEPSY |
|---|
| **Malformation syndromes** |
| Aicardi syndrome |
| Lissencephaly |
| Tuberous sclerosis |
| Sturge-Weber syndrome |
| **Metabolic disorders** |
| Ceroid lipofuscinosis |
| Lafora body disease |
| Phenylketonuria |
| Pyridoxine dependency |
| Tay-Sachs disease |

H. **Convulsion** — seizure associated with violent muscle contractions

I. **Tonic-clonic convulsion** — convulsion characterized by phases of **prolonged muscle contraction** (tonic) and **alternating muscle contraction and relaxation** (clonic)

J. **Absence seizure** — momentary **lapses in consciousness**

K. **Atonic (astatic) seizure (drop attack)** — sudden **loss of tone** in postural muscles

L. **Myoclonic seizure** — synchronous bilateral myoclonic jerks

M. **Todd's paralysis** — **transient** (lasting no more than 48 hours) paralysis during postictal period

| DISORDERS MIMICKING SEIZURES |
|---|
| Breath-holding spell |
| Clonus |
| Complicated migraine |
| Decerebrate/decorticate posturing |
| Drug or alcohol intoxication |
| Hyperventilation |
| Hypoglycemia |
| Narcolepsy |
| Pseudoseizure |
| Sleep myoclonus |
| Syncope |
| Transient ischemic attack (TIA) |

III. **Epileptic phenomena**

A. Sudden (**paroxysmal**) changes in consciousness, sensation, emotion, or thought processes can be manifestations of seizures

B. Must be differentiated from other episodic, periodic, or recurrent paroxysmal non-epileptic events, including syncope, migraine, pseudoseizure, transient ischemic attack (TIA), or narcolepsy

C. **Petit mal** — term frequently mistakenly applied to any minor seizure that falls short of generalized convulsion; however, *petit mal epilepsy* denotes specific clinical syndrome with specific characteristic EEG findings

IV. **Anticonvulsants**

A. Usage — **carbamazepine** (Tegretol), **phenytoin** (Dilantin), **barbiturates** (such as phenobarbital or primidone), **valproate** (Depakote), **gabapentin** (Neurontin), and **lamotrigine** (Lamictal) are effective anticonvulsants, except for absence epilepsy which responds best to **ethosuximide** (Zarontin) or valproate (Depakote)

B. Specific medications

1. **Phenytoin** (Dilantin) — usual adult dosage of 300-400 mg/day; side effects include **gingival hyperplasia**, coarsening of features, **hirsutism**, and exacerbation of **acne**; can cause fatal delayed hypersensitivity reaction (**Stevens-Johnson syndrome**) consisting of fever and pruritic mucocutaneous maculopapular rash with subsequent desquamation

2. **Carbamazepine** (Tegretol) — usual adult dosage of 600 mg/day (in three doses); related structurally to tricyclic antidepressants and can ameliorate behavioral and emotional disturbances accompanying epilepsy; side effects include **bone marrow suppression** and cholestatic **jaundice**

3. **Valproate** (Depakote) — usual adult dosage of 750-1000 mg/day (in three doses); side effects include **liver dysfunction**, **bone marrow depression**, and **pancreatitis**

4. **Phenobarbital** — usual adult dosage of 100-180 mg/day; side effects include exacerbation of **depression**, exacerbation of **behavioral problems in hyperactive children**, and **mental slowing** with long-term use; can cause serious or fatal delayed hypersensitivity reaction (**Stevens-Johnson syndrome**) consisting of fever and pruritic mucocutaneous maculopapular rash with subsequent desquamation

5. **Gabapentin** (Neurontin) — usual adult dosage of 900-1800 mg/day; chemically related to γ-aminobutyric acid (GABA) but does not react with GABA receptors; no significant metabolic interaction with other medications

6. **Lamotrigine** (Lamictal) — usual adult dosage of 150-500 mg/day; also inhibits folate metabolism

7. **Ethosuximide** (Zarontin) — usual dosage of 500 mg/day in older children; side effects include **gastrointestinal upset** and **bone marrow suppression**

V. Electroencephalogram (EEG) — recording of **electrical activity** produced by neurons near cerebral cortical surface; usually performed during interictal period; only records brain activity for finite period of time, thus potentially missing electrical seizure activity; normal EEG does not rule out diagnosis of seizure; techniques to activate electrical seizure activity include **hyperventilation**, photic stimulation, and sleep deprivation; continuous monitoring techniques (video-EEG techniques and ambulatory cassette recording) sample longer periods of time

VI. **Grand mal epilepsy ("classic" grand mal)**

   A. Childhood or adolescent onset of **recurrent generalized (bilaterally symmetric) tonic, clonic, or tonic-clonic seizures**; individual suddenly loses consciousness, body becomes **rigid (tonic phase)** followed by prolonged phase of **jerky movements (clonic phase)**; tongue can be bitten and urinary incontinence can occur; at termination of convulsive activity, individual is confused and sleepy

   B. Interictal EEG is usually normal

   C. Family history of similar seizures is common

   D. Seizures respond to anticonvulsant treatment with phenytoin (Dilantin), barbiturates, or carbamazepine (Tegretol); medication can often be withdrawn at puberty after seizure-free interval with no further recurrence of seizures

VII. **Petit mal epilepsy (absence epilepsy)**

   A. Onset between ages 3 and 10 years of **frequent** (up to hundreds of times per day) **absences** lasting 2 to 30 seconds; associated with characteristic EEG abnormality (during absences) of **3 Hz spike-wave discharges** activated by hyperventilation

   B. **Absences** appear as **staring spells** with momentary unresponsiveness, sometimes associated with eyelid fluttering, eyes rolling upward, lip-smacking, chewing, or slight mouth or hand twitches; absences result in transient pauses during eating, speaking, or other activities and often occur in flurries; absences interfere with school performance and are mistaken for "daydreaming" or "inattentiveness"

   C. More common in females; frequent family history of some type of epilepsy

   D. Must be differentiated from **complex partial (psychomotor, temporal lobe) seizures** which have preceding aura, are of longer duration, and are followed by postictal confusion, drowsiness, or headache

   E. Seizures respond to anticonvulsant treatment with either ethosuximide (Zarontin) or valproate (Depakote); medication can often be withdrawn after seizure-free interval of four years with no further absence seizures; however, some patients have additional tonic-clonic convulsions which require continued medication

   F. **Absence (petit mal) status epilepticus (spike-wave stupor)** — absence seizures occurring nearly continuously, resulting in dazed, blank, or confused appearance; child may intermittently (and slowly) respond to questioning; EEG shows continuous 3 Hz spike-wave discharges; treatment with intravenous diazepam (Valium) stops seizure activity

Chapter 3

VIII. **Benign centrotemporal epilepsy (rolandic epilepsy, sylvian seizure syndrome)**

   A. Syndrome of **partial (focal) seizures** with no evidence of focal structural brain lesion

   B. Onset in elementary school-aged child of **nocturnal secondary generalized convulsions** and rare **daytime partial seizures** characterized by localized tongue or oral paresthesias, dysarthria, drooling, and facial twitching, with preservation of consciousness

   C. EEG shows characteristic runs of central and mid-temporal broad spikes or sharp waves, particularly during sleep; can be distinguished from complex partial (psychomotor; temporal lobe) epilepsy by this characteristic EEG appearance

   D. Family members have similar characteristic EEG pattern inherited as **autosomal dominant trait**; only about 25% of individuals with this EEG pattern develop clinical seizures

   E. Seizures respond to anticonvulsant treatment with phenytoin (Dilantin), barbiturates, or carbamazepine (Tegretol); medication can usually be withdrawn after seizure-free interval of three years with no further recurrence of seizures, although EEG pattern may persist

IX. **Neonatal seizures**

   A. Because of nervous system immaturity, newborns do not have well-organized seizures; instead, seizures take various clinical forms (often all forms can be observed at different times):

   1. **Subtle seizures** — eyelid fluttering or blinking, prolonged (tonic) eye deviation, lip-smacking, drooling, swimming or pedaling movements, apnea

   2. **Tonic seizures** — prolonged (tonic) extension of arms and legs (mimicking decerebrate posturing) or flexion of arms and extension of legs (mimicking decorticate posturing)

   3. **Clonus** — can be multifocal (asymmetric in several limbs) or focal (localized to one limb or part of limb)

   4. **Myoclonus** — synchronous myoclonic jerks

   B. Numerous underlying causes including:

   1. **Early** seizures (birth to age 3 days) — neonatal **hypoxic-ischemic or traumatic brain injury, hypoglycemia** (blood glucose < 30 mg/dL), **intracranial hemorrhage**, or **pyridoxine dependency**

   2. **Later** seizures (after age 3 days) — **infection** (meningitis or encephalitis), hypocalcemia (serum calcium < 7.0 mg/dL), or **inborn errors of metabolism** (such as phenylketonuria, maple syrup urine disease, or galactosemia)

C. Must be differentiated from **neonatal jitteriness (tremulousness)** which can also result from hypoxic-ischemic brain injury, drug withdrawal (following maternal drug addiction), hypocalcemia, or hypoglycemia; jitteriness is **stimulus-sensitive**, consists of rapid regular movements of all extremities that can be abolished by restraint, and does not include abnormal eye movements

D. Following evaluation for possible underlying conditions (search for infection, radiologic imaging studies, blood studies), treatment involves administration of glucose, calcium, and pyridoxine; if seizures continue, phenobarbital or diazepam can be administered; phenytoin can be used intravenously, but it is poorly absorbed with either intramuscular or oral administration

E. Prognosis depends on underlying etiology for seizures

X. **West syndrome (infantile spasms)**

A. Epileptic syndrome in infancy with characteristic triad:

1. **Developmental arrest** — cessation of psychomotor development, with no further acquisition of developmental milestones and loss of previously acquired skills; moderate to severe **mental retardation** in over 90% of these children

2. **Infantile spasms**

a. Bilaterally symmetric seizures lasting less than two seconds consisting of rapid (lightning or Blitzkrämpf seizures) consisting of sudden **massive forward flexion** at neck, waist, and hips with arms extended (termed "jackknife seizures" or "salaam seizures" because of similarity to position of prayer)

b. Seizures occur repetitively (up to hundreds or thousands of times per day), often in flurries particularly upon awakening or when startled or stimulated

3. **Hypsarrhythmia** — disorganized chaotic EEG with irregular high voltage sharp and slow waves

B. Etiologies divided into:

1. **Symptomatic** — slow or abnormal psychomotor development before onset of seizures; associated with **known cerebral disease**, such as following hypoxic-ischemic, infectious, or traumatic brain injury, tuberous sclerosis, Aicardi's syndrome, or phenylketonuria

2. **Cryptogenic** — normally developing infant with **no identifiable brain disease**, who suddenly has seizures and developmental arrest

C. Onset of seizure disorder usually before age 6 months and always before age 12 months; slightly more common in males

D. Treatment

1. Seizures are often **extremely difficult to control**, but in cryptogenic group, early treatment with **adrenocorticotropic hormone (ACTH)** has occasionally produced dramatic and complete remission of seizures (although mental retardation still occurs)

2. In symptomatic group, multiple drugs including prednisone, benzodiazepines, and valproate are often necessary to control seizures; **evolution into Lennox-Gastaut syndrome** commonly occurs

E. Prognosis best for ACTH-responsive infants in cryptogenic group; prognosis for others depends on underlying etiology and ability to control seizures

XI. **Lennox-Gastaut syndrome**

A. Epileptic syndrome (onset between ages 2 to 5 years) with characteristic triad:

1. **Psychomotor retardation**

2. **Multiple seizure types** — atonic seizures (drop attacks), myoclonic seizures, absences, and generalized tonic-clonic convulsions; seizures occur many times per day

3. Characteristic **disorganized EEG** with multifocal slow (less than 3 Hz) spike-wave complexes

B. Etiologies divided into:

1. **Symptomatic** — child with slow or abnormal psychomotor development before onset of seizures and **known cerebral disease**; West syndrome evolving into Lennox-Gastaut syndrome falls into this group

2. **Cryptogenic** — apparently normal development prior to onset of seizures

C. Males affected three times more often than females

D. Treatment with anticonvulsant medications (preferred drug is valproate) only rarely controls seizures

E. Prognosis is poor, with continued seizures and moderate to severe mental retardation

XII. **Benign (uncomplicated, simple) febrile seizures**

   A. Brief (usually less than ten minutes) **generalized (nonfocal) seizure** in child with **fever** prior to onset of seizure and no evidence of central nervous system infection, brain disease, or history of previous seizure without fever; seizure usually occurs during **rapidly rising phase of fever** (thus, generally occurring during first day of febrile illness)

   B. EEG is normal by 2 weeks after seizure

   C. Onset between ages 3 months and 5 years (peak incidence between ages 18 and 24 months); family history of similar febrile seizures is common

   D. **Rarely (less than 2%) develop subsequent epilepsy** with benign febrile seizures and **no other risk factors** (such as known neurologic disease, developmental delay, family history of afebrile seizures, multiple seizures during single illness, or prolonged or focal seizures with fever)

   E. Must be differentiated from seizure with fever (**complicated febrile seizure**)

   1. **Fever lowers seizure threshold** such that fever of any cause can trigger seizures in individuals with **underlying epilepsy**

   2. Fever and seizures can be associated with **central nervous system infection** such as meningitis or encephalitis

   3. Diagnosis of benign febrile seizures can only be assured by thorough examination to **rule out all possible underlying causes** (including central nervous system infection)

   F. Treatment

   1. Acute treatment — seizure has usually ended by time child reaches hospital or clinic; thus, only treatment of febrile illness is necessary

   2. Prophylaxis — argument against medication treatment is that only 30% have second febrile seizure with another febrile illness, only 15% have third febrile seizure, and less than 9% have more than three febrile seizures; thus, medication side-effects must be weighed against low risk of seizure recurrence; anticonvulsants are *not* beneficial if given intermittently (with fevers); prophylaxis with daily phenobarbital or valproate is effective (phenytoin is not), but therapeutic blood levels are necessary

XIII. **Focal motor seizures (simple partial epilepsy)**

   A. **Clonic seizure movements** involving one body part (such as arm, hand, leg, or face); indicates **focal structural lesion** of cerebral **cortical motor strip** controlling that body part

B. Can remain localized, can spread to involve adjacent areas of cerebral cortex with clinical "march" of motor activity to involve other body parts (**jacksonian seizure**), or can become generalized (**secondary generalized seizure**)

C. **Not associated with altered consciousness** unless secondarily generalized; many apparently generalized seizures actually begin as focal motor seizure with rapid spread such that initial focal nature is not observed

D. **Postictal (Todd's) paralysis** common

XIV. **Psychomotor or temporal lobe seizures (complex partial epilepsy)**

A. Initial **aura**

1. **Déjà vu — feeling of familiarity**, as though having experienced event or situation previously

2. **Jamais vu — feeling of unfamiliarity**, as though having never experienced familiar situation or environment

3. **Abdominal or epigastric sensations** (often described as rising sensation)

4. Visual or auditory **hallucinations** — usually complex, vivid and unpleasant

Motor homunculus of left precentral gyrus; specific right-sided body part involved in focal motor seizure depends on site of cortical lesion.

B. **Automatisms — complex, usually coordinated but purposeless motor movements** (such as lip-smacking or walking in circle) with **altered consciousness** (blank stare or stereotyped verbal responses)

C. Can become secondarily generalized with loss of consciousness and tonic-clonic convulsive activity

D. Seizures generally are brief with **postictal** state of **confusion, headache**, and **exhaustion**

E. Temporal lobe spike discharges on routine waking EEG in less than 50% of cases

F. Most common form of epilepsy in adults (40%); frequently associated with interictal personality, behavior, and cognitive disorders resulting in severe psychosocial disability

G. Treatment

1. Anticonvulsant drugs carbamazepine (Tegretol) or phenytoin (Dilantin) reduce seizure frequency in some patients, but many cases are refractory to drug treatment

2. Neurosurgical resection of epileptogenic focus in temporal lobe sometimes leads to remission; histologic study of surgically-removed temporal lobe tissue often reveals structural abnormalities such as hippocampal gliosis (sclerosis), developmental malformation (hamartoma), or low-grade tumor

XV. **Adult-onset seizures** — seizures presenting initially after age 30 years most often are due to either acute progressive or acquired static brain lesions

XVI. **Status epilepticus**

   A. **Continual convulsions** or **recurrent convulsions without regaining consciousness** between seizures

   B. Treatment

      1. Maintenance of adequate ventilation (oxygenation)

      2. Assessment of blood levels of glucose (to rule out hypoglycemia), calcium (to rule out hypocalcemia), and sodium (to rule out hyponatremia)

      3. Administration of glucose intravenously

      4. Intravenous administration of diazepam (Valium) or lorazepam (Ativan) to stop seizures, followed by intravenous phenytoin (Dilantin)

      5. Administration of maintenance doses of anticonvulsant medication

---

**CAUSES OF ADULT-ONSET SEIZURES**
**Toxic-Metabolic**
  Hypocalcemia
  Hypoglycemia
  Hyponatremia
  Intoxication or withdrawal
    Alcohol
    Amphetamine
    Barbiturates
    Cocaine
    Glutethimide
    Meprobamate
    Opiates
    Phencyclidine
  Theophylline toxicity
  Uremia
**Structural**
  Encephalitis
  Meningitis
  Post-infarction
  Post-traumatic
  Tumor

# Chapter 4 — HEADACHE AND CRANIOFACIAL PAIN

I. **Headache** — common complaint; possibly as many as 50% of all patients have headache symptoms; headaches are basically divisible into two broad categories:

   A. **Chronic recurring headaches** — causes include vascular (migraine) headaches, muscular contraction (tension) headaches, or combination of these two types of headache

   B. Headache due to intracranial disease, systemic illness, or local cranial, orbital, or nasopharyngeal disease

II. **Head pain**

   A. Results from sensations originating in skin, subcutaneous tissue, muscles, extracranial and larger intracranial arteries, periosteum (external surface of skull bones), eye, ear, nasopharynx and nasal sinuses, dura along base of brain, cranial nerves (optic, oculomotor, trigeminal, glossopharyngeal, and vagus), and first three cervical nerve roots and branches

   B. Skull bone, dura over convexity of brain, pia-arachnoid, brain parenchyma, ependyma, and choroid plexus all lack sensitivity

   C. Pain originating in extracranial structures tends to be localized, while that originating from intracranial structures tends to be more diffuse

   Anterior location (dark shading) of head pain referred to trigeminal distribution and posterior location (no shading) for upper cervical root distribution.

   1. Pain originating in intracranial **supratentorial** structures is referred to distribution of **trigeminal nerve (cranial nerve V)**

   2. Pain originating in intracranial **posterior fossa** is referred to distribution of **upper cervical roots**

III. **Migraine**

   A. **Periodic recurrent throbbing headaches** of **vascular origin**; **usually unilateral**, but affecting opposite sides during different attacks

   B. Frequently familial; individuals with migraine often have greater than expected prevalence of **colic, motion sickness**, or **episodic abdominal pain** during childhood

C. **Classic migraine** — unilateral throbbing **headache preceded by prodromal aura**

  1. **Aura**

     a. Due to **vasoconstriction** with consequent **cerebral ischemia**

     b. Usually consists of transient (5-30 minutes) **visual symptoms** of scotoma (black or blind spots in vision), unformed light flashes (scintillations), colors, zigzag lines, or hemianopsia (blindness in one-half of visual field)

     c. Rare auras include unilateral numbness or tingling, limb weakness, language difficulties (aphasia), or confusion

  2. **Unilateral throbbing headache** follows aura, increases in severity over several hours, and then slowly resolves after about 24 hours; sleep generally hastens resolution of headache

  3. Irritability, malaise, photophobia, and nausea often associated with headache; sometimes vomiting occurring at peak of headache is associated with rapid resolution of pain

  4. Onset is often in teens; more common in females

D. **Common migraine** — similar to classic migraine, but **lacking prodromal aura**

E. **Complicated migraine**

  1. **Aura symptoms persist** as significant neurologic complications, even after resolution of headache

  2. Must be differentiated from underlying structural lesions (such as arteriovenous malformation) by radiologic imaging studies

  3. Forms include:

     a. **Ophthalmoplegic migraine** — **extraocular muscle palsies** (usually involving oculomotor nerve), generally on same side as headache

     b. **Hemiplegic migraine** — unilateral motor and/or sensory deficits

F. **Basilar artery migraine**

  1. Variant of classical migraine in which aura relates to **ischemia** in distribution of **basilar artery**

2. Aura consists of visual symptoms progressing from blurring and scotoma to blindness, vertigo, ataxia, and **impaired consciousness**; sometimes aura consists of sudden **drop attacks** (loss of postural control) or acute confusional state; aura symptoms can occur without subsequent headache

3. Often presents in early adolescence

G. **Status migrainosis** — daily or virtually **continuous migraine** headaches

H. **Serotonin** and substance P (and possibly other neurotransmitters) implicated in pathogenesis; triggering factors include stress and anxiety, specific foods, alcohol, and weather conditions

I. Treatment is aimed at aborting headache during prodrome phase, symptomatic treatment of pain, or prevention of attacks

1. Ergot derivatives (such as **ergotamine**) alone or in combination with caffeine (Cafergot) can abort headache when taken early in prodrome

2. For most headaches, **aspirin** plus small amount of **barbiturate** sedative (as in Fiorinal) facilitates sleep and subsequent headache resolution, but severe headaches may require additional antiemetics and mild narcotics; for status migrainosis, brief course of **corticosteroids** (prednisone) may also be necessary

3. Various agents have proven useful in preventing attacks, but for any particular individual, one drug may prove more effective than another: **ergotamine, methysergide** (Sansert), **propranolol** (Inderal), **amitriptyline** (Elavil), **cyproheptadine** (Periactin), **phenytoin** (Dilantin), or **verapamil** (Isoptin); cessation of oral contraceptives may also reduce headache frequency in some women

IV. **Cluster headache (Horton's headache, histamine cephalgia, migrainous neuralgia)**

A. Sudden **paroxysmal excruciating unilateral lancinating pain** centered around orbit ("deep behind the eye") and occurring in attacks lasting several minutes to several hours; **clusters of daily attacks** last from three weeks to three months followed by remission for months or years

B. Called "suicide headache" because of severity of pain, which can **awaken** individual from sleep

C. Often associated with unilateral **facial flushing, tearing, nasal stuffiness**, or **partial Horner's syndrome** (ptosis and miosis, but no anhidrosis)

D. About five times more common in men than women; exacerbated by alcohol consumption

E. Treatment is similar to that for migraine; **lithium carbonate** treatment has also been successful in preventing attacks in some individuals

# HEADACHE AND CRANIOFACIAL PAIN

V. **Tension headache (muscle contraction headache)**

    A. **Bilateral, nonthrobbing constant dull aching** headache, often beginning occipitally and diffusely spreading over entire head (sometimes with frontal predominance)

    B. Gradual onset of pain often described as fullness, tightness, or vise-like pressure

    C. Pain is usually **continuous** and may last for days or weeks; some patients experience combination migraine headache (throbbing pain) and tension headache

    D. Treatment

        1. Analgesics (such as aspirin) or narcotics generally provide *no* relief of pain (unlike headaches due to intracranial masses or increased intracranial pressure which are relieved by analgesics)

        2. **Depression** and **anxiety** seem to underlie pathogenesis of headache and require treatment (usually antidepressant pharmacotherapy), which then results in headache remission

VI. **Sinus headache (nasal headache)**

    A. Localized, steady, nonthrobbing pain with **tenderness to percussion** over nasal sinus

    B. Commonly associated with predisposing structural defects (such as deviated septum), allergies, or polyps

    C. Sinus radiographs show mucosal thickening and/or opacification

    D. Treatment includes decongestants, antihistamines, and antibiotics; surgical drainage may occasionally be necessary

VII. **Temporal arteritis (cranial arteritis, giant-cell arteritis)**

    A. **Inflammatory disorder** of cranial arteries in **older individuals** (over age 50 years)

    B. Increasingly severe throbbing or nonthrobbing headache, unilateral or bilateral, often localized over **thickened tender non-pulsatile temporal or other scalp arteries**

    C. Initial **transient visual disturbance** (blurring or blindness) progresses to **permanent blindness** if untreated; for this reason, *temporal arteritis is **medical emergency***

    D. Most patients also have complaints of **malaise, weight-loss,** and generalized **muscle aches** and weakness **(polymyalgia rheumatica)**

E. Characteristically associated with **marked elevation of erythrocyte sedimentation rate** (ESR), often up to 120 mm/hour ("ESR greater than patient's age")

F. Diagnosis established by finding **granulomatous (giant-cell) inflammation** in biopsied temporal artery segment

G. **Immediate high-dose corticosteroids** are necessary to prevent permanent visual loss and in suspected cases must be initiated prior to final histopathologic confirmation; corticosteroid dosage can be reduced when ESR returns to normal and since disease is generally self-limited, medication can be discontinued after 6 months to 2 years

VIII. **Trigeminal neuralgia (tic douloureux)**

A. Brief **paroxysmal lancinating pain** in distribution of **trigeminal nerve** (cranial nerve V) causing patient to grimace (tic); second and third divisions more commonly affected than first division

B. Paroxysms can be initiated by touching or moving **trigger points** on face

C. No identifiable weakness or sensory loss

D. Onset usually after age 40 years; women more commonly afflicted than men; etiology unknown

E. Must be differentiated from atypical trigeminal neuralgia due to multiple sclerosis or posterior fossa mass lesion (tumor or aneurysm), both of which usually present in younger individuals and have associated sensory loss in trigeminal distribution or other cranial nerve palsies

F. Treatment with **carbamazepine** relieves symptoms in most cases, but refractory cases may require surgical ablation of gasserian (trigeminal) ganglion or vascular decompression of trigeminal nerve root entry zone

IX. **Post-lumbar puncture headache**

A. Following lumbar puncture, frontal, occipital, or generalized headache is **precipitated by sitting or standing** and **relieved promptly by lying flat**

B. Due to **cerebrospinal fluid leakage** through dural hole produced by spinal needle; more common with larger bore spinal needles, multiple punctures, or poor lumbar puncture technique

C. Treatment involves strict bed rest and adequate hydration

X. **Postconcussion syndrome**

A. **Headache, dizziness, and personality changes** following concussive head injury

## HEADACHE AND CRANIOFACIAL PAIN

B. Usually associated with **anxiety** and **depression**, but must be differentiated from intracranial lesions (such as subdural hematoma) by performing radiologic imaging studies

C. Treatment involves reassurance, analgesics, or antidepressant medications; if litigation is pending, therapeutic measures are usually ineffective

XI. **Pseudotumor cerebri (benign intracranial hypertension)**

A. Syndrome of **increased intracranial pressure** (without mass lesion or hydrocephalus), **headache**, and **papilledema**

B. Women affected eight times more frequently than men; typically occurs in **obese women** of child-bearing age

C. Dull, intermittent or steady, **nonthrobbing headache**, often **worse in morning** and during Valsalva maneuver or other straining efforts such as coughing, sneezing, or bending forward

D. **Blurred vision** and transient visual obscurations progress to **constriction of visual fields** and **enlarged blind spots**; if untreated, permanent visual loss ensues

E. Occasionally associated with complaints of diplopia (from abducens nerve damage) and facial tingling or numbness

F. Radiologic imaging studies demonstrate **small slit-like ventricles** and lumbar puncture shows normal cerebrospinal fluid under **markedly elevated pressure** (250-500 mm of water)

G. Treatment involves **weight reduction**, and repeated lumbar punctures to remove fluid and lower pressure; administration of acetazolamide (Diamox) or corticosteroids can reduce cerebrospinal production; progression of visual loss necessitates surgical treatment such as optic nerve sheath decompression or insertion of lumbo-peritoneal cerebrospinal fluid shunt

XII. **Headache from meningeal irritation**

A. Headache in association with **fever** and **stiff neck** suggests **meningitis** or **encephalitis**

B. **Sudden onset** of severe headache and **stiff neck** suggests **subarachnoid hemorrhage** from leaking **arteriovenous malformation**, **ruptured saccular (berry) aneurysm**, or intraparenchymal hemorrhage

# Chapter 5          TOXIC, METABOLIC, AND NUTRITIONAL DISEASES

I. Many neurologic disorders are subsumed under this designation and some disorders previously considered to be neurodegenerative are now known to be due to metabolic abnormality

II. **Alcohol (ethanol) poisoning** — most common nervous system toxin with wide variety of effects related to acute ingestion, chronic exposure, or withdrawal

    A. **Metabolism**

        1. Alcohol is rapidly and completely absorbed from gastrointestinal tract, initially appearing in blood 5 minutes after ingestion and reaching peak blood levels within 90 minutes

        2. **Oxidatively metabolized in liver** at rate of approximately 0.15 grams of alcohol per kilogram body weight per hour (about 10 grams/hour in adult male) which is equivalent to alcohol content in 1 ounce of 90-proof whiskey or 10 ounces of beer

| SYMPTOMS AND BLOOD ALCOHL LEVEL | |
| --- | --- |
| 0.3 - 0.5 g/L | euphoria or dysphoria |
| 0.5 - 1.0 g/L | incoordination, impaired judgment, inattentiveness |
| 1.0 - 2.0 g/L | ataxia, slurred speech, labile mood |
| 2.0 - 3.0 g/L | confusion, drowsiness |
| 3.0 - 4.0 g/L | stupor |
| 4.0 - 5.0 g/L | coma |

    B. Acute toxicity (**alcohol intoxication**)

        1. Symptoms of acute alcohol ingestion vary with blood alcohol level; lethal blood alcohol level is approximately 5 g/L

        2. Blood alcohol level of 2.0 g/L, without overt clinical symptoms of intoxication suggests alcohol tolerance and implies chronic (heavy and persistent) alcohol ingestion

    C. **Hangover — headache, nausea, vomiting, irritability**, and **tremulousness** that occur several hours after brief, excessive alcohol consumption; does not imply alcohol abuse

    D. **Chronic alcohol ingestion (alcohol abuse; alcoholism)**

        1. **Tolerance** develops following chronic alcohol exposure due to various effects on neurons including alteration of membrane receptor systems

        2. Alcohol-induced alterations in intracellular metabolism, systemic organ dysfunction (particularly liver, kidney, and gastrointestinal tract), and nutritional disturbances all adversely affect brain function

# TOXIC, METABOLIC, AND NUTRITIONAL DISEASES

3. Diagnosis of **alcohol abuse**

   a. **Pattern of need for daily alcohol** use to maintain adequate social functioning, **inability to reduce or stop** alcohol consumption despite repeated efforts (including periods of temporary abstinence), **binges** of heavy alcohol consumption, **amnesia** for events occurring while intoxicated (**alcoholic "blackouts"**), continuing alcohol consumption despite physical disorders exacerbated by alcohol use, or drinking of non-beverage alcohol (such as vanilla extract)

   b. Quantity of alcohol consumed does not necessarily define alcoholism since body weight, nutrition, ethnic (genetic) background, and general health all influence sensitivity to acute and chronic effects of alcohol

   c. Clues to alcohol abuse include evidence of trauma incurred during periods of drunkenness, poor nutrition, unexplained persistent or recurrent infections, red blood cell count revealing mean corpuscular volume (MCV) greater than 97 fL with round macrocytosis, or liver disease (or elevated serum levels of liver enzymes)

E. **Alcohol-induced coma** — requires blood alcohol level of greater than 3.0 g/L and **no other cause for coma** (such as subdural hematoma, hypoglycemia, hypoxia, or meningitis)

F. **Alcoholic "blackouts"**

   1. Transient **amnesia** for periods of alcohol intoxication, despite observers noting no alteration in consciousness during these periods

   2. **Mimics transient global amnesia**, which is form of **transient ischemic attack** characterized by brief period of bewilderment in which (usually elderly) individual has defective memory for events of present and recent past, but no alteration in consciousness and no abnormal neurologic signs

G. **Alcohol withdrawal syndrome (abstinence in chronic alcoholism)**

   1. **Alcohol abstinence** (relative or complete) that produces **falling blood alcohol level** results in symptoms that can develop alone or together

      a. **Early** — tremulousness, alcoholic hallucinosis, and **withdrawal seizures**

      b. **Late** — **delirium tremens**

   2. Withdrawal symptoms can occur as late as seven days after cessation of alcohol intake, but prolonged or delayed withdrawal symptoms suggest possibility of abuse of other drugs along with alcohol (can also occur in patients receiving tranquilizers or sedatives as part of alcohol treatment program)

3. **Tremulousness**

    a. Most common withdrawal symptom, usually appearing in morning following overnight abstinence

    b. **Generalized tremor** (referred to as "the shakes" or "the jitters") and associated with **irritability, jerky movements, hyperalertness, insomnia, tachycardia, gastrointestinal upset** (nausea or vomiting), **facial flushing**, and **diaphoresis**

    c. If abstinence continues, symptoms improve over several days but can persist for up to two weeks; however, symptoms improve immediately following "a few drinks to quiet my nerves"

4. **Alcoholic hallucinosis**

    a. Occurs in about 25% of individuals with tremulousness

    b. **Perceptual disturbances** consisting of **terrifying nightmares** (related to **rebound of REM sleep after alcohol-induced REM suppression**) and **fragmentary waking hallucinations** (most often visual)

    c. Repeated acute auditory hallucinations can evolve into persistent **chronic auditory hallucinations** with paranoid delusions resembling schizophrenia

5. **Withdrawal seizures ("rum fits")**

    a. Flurry of several **brief generalized tonic-clonic convulsions** with loss of consciousness, occurring **within 48 hours** after ceasing alcohol consumption; can be triggered by stroboscopic stimulation (such as flickering lights)

    b. **Alcohol can precipitate seizures in known epileptics**, particularly after brief period (several hours) of abstinence (such as in morning after overnight abstinence); additionally, cerebral cortical scars (from repeated head trauma) are common in alcoholics and may trigger seizures during alcohol withdrawal

    c. Diagnosis of withdrawal seizures can only be made if seizure is generalized (**not focal**) and has typical clinical pattern, no seizures unassociated with alcohol have occurred, and there is **no other identifiable cause** to account for seizures (such as hypoglycemia, meningitis, subdural hematoma, or pre-existing epilepsy)

    d. Withdrawal seizures do not recur and **do not predispose to epilepsy**, if individual remains abstinent

# TOXIC, METABOLIC, AND NUTRITIONAL DISEASES

      e.    About one-third of patients with **alcohol withdrawal seizures** subsequently develop **delirium tremens**

  6.  **Delirium tremens**

      a.    Rapid onset of **profound confusion, agitation, tremor, delusions, extremely vivid visual hallucinations, insomnia**, and **autonomic nervous system hyperactivity (fever, tachycardia, dilated pupils**, and **profuse sweating)**

      b.    Onset 48 to 96 hours following cessation of alcohol consumption, with peak incidence after 72 hours

      c.    Individual appears **restless and distracted**, often carrying on **imaginary conversations** or seemingly trying to fend off **hallucinated** people or objects

      d.    Mortality can be as high as 15%, mostly due to circulatory collapse

  7.  Diagnosis of withdrawal syndrome is not exclusionary and other potential complications of alcohol abuse (such as subdural hematoma, thiamine deficiency, hypoglycemia, electrolyte disturbances, anemia, pneumonia, pancreatitis, liver failure, or meningitis) must be identified and treated

  8.  Treatment

      a.    Prevention or reduction of **early symptoms** can be achieved with administration of **paraldehyde** or **benzodiazepines**, such as diazepam (Valium) or chlordiazepoxide (Librium), for several days followed by tapering of dosage; **ß-adrenergic blocking drugs**, such as propranolol (Inderal), can decrease heart rate and reduce tremor; withdrawal seizures usually end before treatment with anticonvulsant medications can be instituted or would become effective

      b.    Alcoholic individual with epilepsy who continues to consume alcoholic beverages should not be given anticonvulsants, since compliance will be poor

      c.    Treatment of **delirium tremens** requires **volume replacement** and careful maintenance of **fluid and electrolyte balance**; sedation is achieved by parenteral administration of benzodiazepines, such as diazepam (Valium) or chlordiazepoxide (Librium)

H.  **Wernicke-Korsakoff syndrome**

  1.  **Wernicke disease (encephalopathy)** — abrupt onset of **nystagmus** and **lateral rectus palsy** ("cross eyes") progressing to **external ophthalmoplegia** (paralysis of eye movement) associated with truncal and gait **ataxia** and **global confusional state**; while **ocular palsies**

and ataxia improve, confusional state transforms into **amnesic syndrome of Korsakoff's psychosis**

2. **Korsakoff's psychosis** — amnesic syndrome following Wernicke disease characterized by **inability to form new memories (anterograde amnesia)** despite relatively intact immediate recall and relatively preserved remote memory; also characterized by **confabulation** (tendency to **falsify memories** by **"filling in" gaps in memory** with information that sounds plausible but has little basis in reality); alertness, attentiveness, and other behavioral functions are normal

3. Pathogenesis is acute **thiamine deficiency**; treatment with **immediate high-dose parenteral thiamine** during *early* Wernicke disease will produce rapid dramatic improvement in ocular palsies with slower resolution of ataxia and confusional state; **delayed administration of thiamine** (presumably after neuronal destruction has occurred) results in **residual ataxia and mental disturbances (Korsakoff's psychosis)**

4. Pathologic changes consist of symmetrical **necrosis of mammillary bodies** along with necrosis of gray matter adjacent to third ventricle, aqueduct, and floor of fourth ventricle; individuals with **Korsakoff's psychosis** also have prominent destruction of **dorsomedial nucleus of thalamus**

I. **Alcoholic cerebellar degeneration**

1. Slowly progressive (over weeks or months) **truncal and gait ataxia**, characterized by incoordination of leg movements and truncal instability

2. Characteristic pathologic change (often visible on radiologic imaging studies) is **degeneration of anterior superior cerebellar vermis (anterior lobe cerebellar atrophy)**

3. Presumably related to nutritional deficiency, but only some patients seem to improve following thiamine supplementation

J. **Alcoholic polyneuropathy — slowly progressive sensorimotor polyneuropathy** presenting with hypoactive or absent ankle reflexes, complaints of **burning or painful feet**, and excessive sweating of feet and hands; presumably related to thiamine deficiency in combination with deficiency of other B vitamins

K. **Alcoholic cerebral atrophy** — radiologic imaging studies (particularly MRI) performed in alcoholics frequently demonstrate dilated ventricles and widened cerebral sulci; however, after sustained abstinence, repeat imaging studies often reveal normal ventricular and sulci size, suggesting **reversible** alcohol-induced shifts in brain fluid content rather than loss of tissue

# TOXIC, METABOLIC, AND NUTRITIONAL DISEASES

III. **Methanol (methyl alcohol; wood alcohol) poisoning**

 A. Ingestion of as little as 60 mL can be fatal; commonly found as adulterant in alcoholic beverages (particularly illegally-produced beverages such as "moonshine")

 B. Initial symptoms of drunkenness, headache, abdominal pain, and visual loss, evolve into delirium and coma; **acidosis** can be severe

 C. If recovery occurs, **blindness** often persists due to retinal and optic nerve damage

IV. **Ethylene glycol poisoning**

 A. Major constituent of antifreeze, which can be inadvertently consumed by alcoholics; ingestion of 120 mL often is fatal

 B. Initial presentation of drunkenness, followed by **generalized convulsions** and coma

 C. Metabolic conversion into oxalic acid produces severe **acidosis** and **oxalate crystal deposition** in **kidneys** (resulting in **uremia**) and in **brain** (resulting in **chemical meningitis** with increased numbers of cerebrospinal fluid lymphocytes)

V. **Opiate abuse**

 A. **Overdosage**

 1. Although considerable tolerance develops to effects of opiates and opiate analogues, overdosage in addicts can still occur accidentally or from suicide attempts

 2. Initial presentation of **stupor or coma, slow shallow breathing** (often with cyanosis), **pinpoint pupils, bradycardia** (often with hypotension), and **hypothermia**

 3. Treatment involves emergency **ventilatory support** and **intravenous administration of opiate antagonist naloxone** (Narcan)

 B. **Withdrawal syndrome**

 1. Following **abstinence** (interval of time varies depending on half-life of particular addicting drug), onset of progressive **yawning, rhinorrhea, profuse sweating, insomnia, pupillary dilation, restlessness, muscular twitching,** complaints of "**hot and cold flashes**" and severe **muscle aches, nausea** and **vomiting, diarrhea, tachypnea, tachycardia,** and **elevated blood pressure**; symptoms peak after several hours or days and then slowly subside; dehydration and electrolyte imbalance can occur without sufficient fluid replacement

Chapter 5

        2.    Centrally-acting $\alpha_2$-adrenergic receptor agonist clonidine (Catapres) alleviates symptoms of opiate withdrawal syndrome

VI.    **Psychostimulant toxicity** — **amphetamines** and **cocaine** produce similar acute toxic effects of **psychosis, hyperthermia, hypertension** (including hypertensive **intracranial hemorrhage**), **pupillary dilation, cardiac arrhythmias, convulsions, coma**, respiratory arrest, and death

VII.   **Lead poisoning**

     A.   **Acute encephalopathy** — infants or young children with **pica** (eating of non-nutritive substances such as dirt, clay, or flaking paint containing lead) present with **irritability, lethargy, ataxia, seizures**, and **coma**; often fatal due to **massive cerebral edema** and diffuse neuronal necrosis; survivors have residual mental retardation, seizures, ataxia, and spasticity

     B.   **Motor neuropathy** — chronic lead exposure in adults mimics mononeuritis multiplex (commonly **wrist drop** or **foot drop**)

     C.   Associated with **anemia, basophilic stippling** of red blood cells, **"lead line"** along gingiva, **constipation**, and **colicky abdominal pain**

VIII.  **Arsenic poisoning**

     A.   **Neuropathy** — **painful sensorimotor neuropathy** (red burning hands and feet)

     B.   **Encephalopathy** — progressive fatigue, lethargy, headache, confusion, seizures, coma, and death; autopsy reveals punctate **white matter hemorrhages** ("brain purpura")

     C.   Associated with **anemia, brown skin discoloration**, plantar and palmar **hyperkeratosis**, and white transverse bands in nails (**Mees' lines**); increased levels of arsenic can be detected in hair and urine

IX.   **Mercury poisoning**

     A.   **Adults** — **dementia** ("Mad Hatter syndrome"), along with cerebellar ataxia, intention tremor, and **motor neuropathy**

     B.   **Children** — **acrodynia** (Pink disease) characterized by swollen, red, cold, moist hands and feet, along with irritability, insomnia, and anorexia

     C.   Organic mercury can cross placenta causing mental retardation or cerebral palsy

X.    Vitamin A excess (**hypervitaminosis A**) — associated with **pseudotumor cerebri**, which resolves with cessation of vitamin A intake

XI. **Vitamin E (α-tocopherol) deficiency** — chronic **malabsorption syndromes** (such as occurs with cystic fibrosis, abetalipoproteinemia, liver disease, or intestinal resections) often result in deficiency of fat-soluble vitamins (vitamins A, D, K, and E); deficiency of vitamin E results in **peripheral polyneuropathy** and **ataxia** mimicking spinocerebellar degeneration; treatment with vitamin supplementation improves symptoms

XII. **Heat stroke**

   A. In hot environment, failure of central nervous system regulatory mechanisms controlling heat loss results in rapid rise in body temperature above 41°C; skin is hot and dry

   B. Relatively rapid alteration of consciousness progressing from confusion to **stupor or coma** (**seizures** are common) with evidence of **increased intracranial pressure** due to generalized **cerebral edema**; associated systemic disturbances include **dehydration**, disseminated intravascular coagulation, and circulatory collapse

   C. Treatment involves **immediate cooling** of body surface with ice water, core cooling with cold gastric lavage and cool intravenous fluids, ventilatory support, reduction of increased intracranial pressure (with hypertonic solutions such as mannitol), and administration of anticonvulsant medications to stop seizures

   D. Survivors have varying neurologic signs based upon extent of neuronal degeneration; often greatest damage is to cerebellar **Purkinje cells**, which can be completely destroyed

XIII. **Hepatic encephalopathy**

   A. Following gastrointestinal hemorrhage or other insult causing **elevation of blood ammonia levels** in patient with **chronic liver disease** and portosystemic shunting (from cirrhosis and portal hypertension), development of progressive **ataxia, dysarthria, asterixis**, lethargy, stupor, and finally coma; characteristic EEG abnormalities can be detected

   B. Reduction in blood ammonia level results in clinical improvement, but symptoms return with further elevation of blood ammonia level; after repeated episodes of hepatic encephalopathy with recovery, **chronic dementia** develops

XIV. **Uremia**

   A. **Encephalopathy**

      1. Progressive **lethargy**, stupor, and **coma** associated with renal failure; reduced level of consciousness correlates with elevated blood urea nitrogen (BUN) level

      2. **Seizures** are common and can relate to several mechanisms: uremia, hyponatremia, hypocalcemia, hypomagnesemia, or hypertensive encephalopathy

B. **Neuropathy**

1. **Painful** distal sensorimotor polyneuropathy; often described as "burning feet"

2. Does not correlate with degree of uremia; usually only responsive to restoration of normal renal function (renal transplantation)

XV. **Carbon monoxide poisoning**

A. **Hypoxia without cyanosis; "cherry-red" appearance**

B. Initial lethargy progresses to coma; severe hypoxia results in brain stem and myocardial dysfunction usually leading to death

C. Treatment by early removal from carbon monoxide source and administration of 100% oxygen can result in survival; survivors often have residual neurologic signs from neuronal destruction, particularly **basal ganglia damage** resulting in choreoathetosis or parkinsonism

XVI. **Hypoglycemic encephalopathy**

A. Brain dysfunction related to critically **low blood glucose** levels

1. Blood glucose of 30 mg/dL is associated with **headache, tremulousness, hunger, diaphoresis, confusion**, and **drowsiness**; persistent low blood glucose level results in coma, decerebrate rigidity, and seizures

2. Blood glucose of 10 mg/dL is associated with profound coma, pallor, shallow breathing, bradycardia, and hypotonia; termed "medullary phase of hypoglycemia" and suggests cerebral neuronal death

B. Treatment requires **immediate intravenous glucose** administration (usually bolus of 50% glucose solution); early treatment can result in nearly complete recovery, but delayed treatment is associated with cerebral cortical neuronal destruction and residual neurologic signs

XVII. **Kernicterus**

A. **Unconjugated bilirubin** passes through **immature blood-brain barrier** compromised by hypoxia and acidosis; **hyperbilirubinemia** results in varying degrees of **bilirubin deposition** in basal ganglia, thalamus, hippocampus, inferior olivary nuclei, subthalamic nuclei, dentate nuclei, and most **cranial nerve nuclei**; bilirubin deposition produces neuronal death and reactive glial scarring

## TOXIC, METABOLIC, AND NUTRITIONAL DISEASES

   B. Associated with serum bilirubin levels greater than 20 mg/dL in full term neonates (usually from **erythroblastosis fetalis** due to Rh or ABO incompatibility); much lower serum bilirubin levels (10 mg/dL) can result in bilirubin deposition in sick or premature newborns

   C. Infants surviving neonatal period have **deafness** and **choreoathetoid cerebral palsy**

XVIII. **Acute intermittent porphyria**

   A. **Autosomal dominant disorder** of porphyrin metabolism due to deficiency of enzyme porphobilinogen deaminase (uroporphyrinogen I synthase) resulting in elevated urinary excretion of **δ-aminolevulinic acid** and **porphobilinogen** (measured with **Watson-Schwartz test**)

   B. Characterized by **recurrent acute attacks**

   1. Initial symptoms — **abdominal pain**, vomiting, and constipation with associated fever and leukocytosis; sometimes mistaken for acute appendicitis or acute abdominal crisis

   2. Neurologic symptoms — cerebral symptoms of **psychosis**, confusion, delirium, and occasionally **seizures**, along with peripheral and cranial motor (and sometimes autonomic) **neuropathy**; differentiated from Guillain-Barré syndrome by normal cerebrospinal fluid protein level

   3. Recovery often is complete, although some patients can have residual weakness

   C. Treatment of acute attacks consists of intravenous administration of **hematin** which inhibits porphyrin synthesis; **high-carbohydrate diet** tends to suppress attacks

   D. Attacks can be precipitated by drug administration or metabolic changes that **induce porphyrin biosynthesis**:

   1. Various drugs — including **barbiturates** and other anticonvulsants, **sedative-hypnotics**, **sulfonamides**, **alcohol**, and chloramphenicol

   2. Fluctuations in levels of sex steroid hormones (estrogen, progesterone, and testosterone)

   3. **Starvation** or **low carbohydrate intake** (as in dieting or with illness)

**Chapter 6**                  DEMENTIA AND NEURODEGENERATIVE DISEASES

I. **Dementia** — general term for deterioration of intellectual/cognitive abilities of sufficient severity to interfere with normal social functioning; dementia can occur at any age, but term is often used to imply neurodegenerative disease affecting older individuals

Lightly stroking palm causes patient to grasp.

A. Signs of dementia include **confusion, memory disturbances, difficulties with problem solving and abstract thinking**, and **impaired judgment**; personality changes, emotional lability, and irritability also may be evident

B. Findings of **frontal lobe dysfunction** often associated with dementia include:

1. **Bruns' ataxia** — **broad-based, flat-footed** gait with **short steps** and tendency to **retropulsion**

2. **Grasp reflex** — **grasp** elicited by lightly stroking palm

3. **Rooting reflex** — **lips turn toward stroking stimulus** at corner of mouth

4. **Snout reflex** — **puckering of lips** in response to gentle tapping of upper lip

With rooting reflex, lips turn toward stroking stimulus at corner of mouth.

5. **Paratonic rigidity (gegenhalten)** — passive manipulation of limbs results in **resistance to all movement**

6. **Perseveration** — inappropriate **repetition** of activity

7. **Impersistence** — inability to persist in task

C. Must be differentiated from acute confusional state or **delirium** (acute or subacute onset of cognitive/intellectual impairment with attentional disorder and altered level of consciousness)

II. **Treatable dementia** — specific treatable disorders (which must be differentiated from neurodegenerative dementia) including:

A. **Normal pressure hydrocephalus** — triad of **unsteady gait, dementia** (particularly memory deficit and psychomotor retardation), and **urinary incontinence** associated with CT scan

# DEMENTIA AND NEURODEGENERATIVE DISEASES

evidence of **large ventricles**; finding of symptomatic improvement following lumbar puncture suggests favorable response to ventricular shunt insertion

B. **Depressive pseudodementia**

1. Refers to **major depressive disorder** in elderly individuals with cognitive disturbance

2. Vegetative symptoms of depression (sleep disturbance, psychomotor retardation, lack of energy) are often prominent

3. Treatment involves antidepressant pharmacotherapy or electroconvulsive therapy

III. **Vascular dementia**

A. Second most common cause of dementia (behind Alzheimer's disease)

1. Most patients have history of **hypertension**

2. Other risk factors include stenosis or occlusion of cervical carotid arteries or emboli of cardiac origin

B. Term is frequently misused as explanation for many causes of dementia; requires specific evidence that brain infarction explains cognitive deficit (since many treatable or neurodegenerative dementias affect older individuals at high risk for arteriosclerotic cerebrovascular disease, incidental infarction can be found in many such patients; yet vascular lesions do not necessarily explain progressive dementia)

C. **Multi-infarct dementia** — careful history and examination indicate a **"stair step" progression of symptoms** (ie, periods of stability interspersed with sudden deterioration) associated with **multiple identifiable cerebral cortical infarcts**

D. **Lacunar state** — **multiple small lacunar strokes (cavities)** in basal ganglia, thalamus, and periventricular white matter

E. **Binswanger's disease** — lateral **ventricular enlargement** associated with **demyelination** of cerebral hemispheric white matter, but sparing short subcortical association fibers (U-fibers)

---

**SELECTED CAUSES OF DELIRIUM**

**Brain disease**
  Brain tumor
  Encephalitis
  Meningitis
  Subdural hematoma
**Cardiac disease**
  Arrhythmias
  Congestive heart failure
  Hypotension
**Drug intoxication**
**Infections**
**Metabolic disorders**
  Anemia
  Azotemia/renal failure
  Fever
  Hyponatremia
  Hypothermia
  Hypovolemia
**Postsurgical/postanesthesia**
**Pulmonary disease**
  Hypoxia/hypercarbia
**Sensory deprivation**
  Blindness/deafness
**Severe pain**
  Fecal impaction
  Urinary retention
**Sleep deprivation**

Chapter 6

IV.  Neurodegenerative dementia

   A.  **Alzheimer's disease**

      1. Most common cause of dementia, accounting for over 65% of cases of dementia over age 65 years and for nearly 50% of nursing home admissions; incidence increases with age

      2. Diagnosis of *probable* Alzheimer's disease made on basis of **clinical evidence of progressive dementia** with **no disturbance of consciousness** and **absence of systemic or other brain diseases** that cause dementia

      3. Diagnosis of *definite* Alzheimer's disease requires **neuropathologic (biopsy or autopsy) confirmation** of clinical diagnosis

      4. Neuropathologic findings include nonspecific generalized cerebral cortical atrophy and ventricular enlargement; diagnostic microscopic abnormalities are:

         a. **Neurofibrillary tangles** — clumps of **paired helical filaments** occupying neuronal cell body and axonal and dendritic processes

         b. **Neuritic (senile) plaques**

            (1) Core of extracellular **amyloid** (ß/A4-amyloid) produced by abnormal metabolism of membrane protein (amyloid precursor protein)

            (2) Periphery of abnormal neurites (nerve cell processes containing distended lysosomes and paired helical filaments), microglia (brain macrophages), and reactive astrocytes

      5. Marked **loss of cholinergic neurons** of nucleus basalis of Meynert (which send processes to hippocampus) may explain early memory disturbances

      6. Genetic factors involved

         a. **Autosomal dominant inheritance pattern in some families**

---

**SELECTED TREATABLE DEMENTIAS**

**Affective disorder (depression)**
**Autoimmune disease**
  Multiple sclerosis
  Systemic lupus erythematosus
**Infections**
  Neurosyphilis
  Fungal meningitis
  Parasitic infestation
  Whipple's disease
**Metabolic disorders**
  Cushing's syndrome
  Hypopituitarism
  Liver failure (hepatic encephalopathy)
  Porphyria
  Renal failure (uremia)
  Hypercalcemia
  Hypoglycemia
  Hypothyroidism
**Normal pressure hydrocephalus**
**Nutritional deficiency**
  Folate deficiency
  Thiamine deficiency (Wernicke-Korsakoff)
  Vitamin $B_6$ deficiency (pellagra)
  Vitamin $B_{12}$ deficiency
**Space-occupying lesions**
  Brain tumor
  Subdural hematoma
**Toxicity**
  Alcoholism
  Drug toxicity
  Heavy metal poisoning
  Organic chemical poisoning

# DEMENTIA AND NEURODEGENERATIVE DISEASES

      b.    Loci on chromosomes 1, 14, 19, and 21 associated with genetic predisposition

           (1)  Mutations in amyloid precursor protein (chromosome 21) predispose to disease in younger individuals

           (2)  Apolipoprotein E $\epsilon$4 phenotype (chromosome 19) predisposes to disease in elderly individuals

      c.    **Clinical dementia** with **histopathologic findings of Alzheimer's disease** occurs in virtually all patients with **Down syndrome** (trisomy 21) living beyond age 30 years

  7.  Treatment involves supportive care for patient and family; psychotropic (neuroleptic) drugs can be used to control severe agitation; cholinomimetic drugs such as tacrine (tetrahydroacridinamine or THA; Cognex) can improve memory function in some patients; growth factors that may enhance neuronal survival have mainly investigational uses currently; benefits from nootropic agent hydergine have not been proven

---

**SELECTED DEGENERATIVE DEMENTIAS**

**Neurodegenerative disorders**
   Alzheimer's disease
   Diffuse Lewy body disease
   Huntington's disease
   Parkinson's disease
   Parkinsonism-ALS-dementia of Guam
   Pick's disease
   Progressive supranuclear palsy
   Spinocerebellar degeneration

**Traumatic dementia**
   Dementia pugilistica ("punch drunk")

**Infectious dementia**
   AIDS encephalopathy
   Behçet's disease
   Creutzfeldt-Jakob disease
   Progressive multifocal leukoencephalopathy

---

B.  **Pick's disease**

  1.  **Progressive dementia** with early **prominent alterations in emotion, affect, and personality**; often clinically indistinguishable from Alzheimer's disease

  2.  Characteristic neuropathologic feature of prominent **frontotemporal atrophy** with sparing of posterior portion of superior temporal gyrus; pattern occasionally visible on CT scan or MRI during life

  3.  Histopathologic finding of characteristic neuronal inclusions termed **Pick bodies**

  4.  Many cases have familial clustering suggestive of autosomal dominant inheritance

V.  **Infectious dementia**

    A.  **Creutzfeldt-Jakob disease** (transmissible spongiform encephalopathy)

1. **Transmissible encephalopathy** related to kuru (disease of New Guinea highland natives), sheep scrapie, transmissible mink encephalopathy, bovine spongiform encephalopathy ("mad cow disease")

2. **Rapidly progressive dementia** associated with **ataxia** and **myoclonus**

3. Electroencephalogram shows high voltage slow-sharp wave complexes on nearly flat background ("**burst suppression**" **pattern**)

4. Causative agent composed of protein only (no nucleic acids) termed **prion**; resistant to standard sterilization procedures (boiling, formalin, alcohol, or ultraviolet radiation); inactivated by autoclaving or immersion in sodium hypochlorite (household bleach) or sodium hydroxide

5. Transmissibility possible from neural tissues; **iatrogenic transmission** demonstrated after corneal transplantation, implantation in brain of infected recording electrodes, and injections of growth hormone extracted from cadaveric human pituitary; thus, **dementia patients should not be used as organ donors**

B. **AIDS (acquired immunodeficiency syndrome) dementia**

1. **Progressive dementia** developing during late stages of **human immunodeficiency virus (HIV)** infection

2. Accompanied early by **apathy, ataxia, disturbances of ocular motility,** and **hyperactive tendon reflexes**, with later **paraplegia** and **incontinence**

3. Pathologic change is rarefaction of white matter with perivascular collections of microglia, foamy macrophages, and **multinucleated giant cells**; HIV can be demonstrated in perivascular cells

4. Can be **complicated by opportunistic infections** (including **cryptococcal meningitis, tuberculous meningitis, neurosyphilis, cerebral toxoplasmosis, aspergillus** emboli with infarction, **cytomegalovirus encephalitis,** or **progressive multifocal leukoencephalopathy**) and malignancy (primary central nervous system **lymphoma** or metastatic tumors)

C. **Progressive multifocal leukoencephalopathy**

1. **Progressive neurologic disorder** occurring in setting of **immunologic compromise** associated with lymphoreticular malignancies (such as lymphoma or chronic lymphocytic leukemia), transplant immunosuppression, AIDS, or immunosuppressive drug treatment

## DEMENTIA AND NEURODEGENERATIVE DISEASES

2. **Rapid evolution of dementia, ataxia, visual field defects, spasticity and weakness** (from corticospinal tract damage), and **swallowing and speech difficulties**, followed by coma and **death usually within six months** of onset; no effective treatment is known

3. Pathologic change is **widespread destruction of white matter**; microscopic examination shows **giant astrocytes with bizarre-shaped nuclei** and enlarged oligodendrocytes with inclusions

4. Causative agent is an opportunistic **human polyoma virus, JC virus**

VI. **Spinocerebellar degeneration** — heterogeneous group of **heredofamilial** disorders characterized by prominent **ataxia**

   A. **Friedreich's ataxia**

   1. Onset of **progressive limb and truncal ataxia** before age 20 years

   2. Additional symptoms include dysarthria, **areflexia, spasticity** and extensor plantar response (Babinski reflex), **pes cavus foot deformity, kyphoscoliosis**, and **cardiomyopathy**; death from cardiac and pulmonary complications usually occurs in third or fourth decade of life

   3. **Autosomal recessive inheritance** due to gene abnormality linked to chromosome 9

   B. **Olivopontocerebellar atrophy**

   1. **Progressive ataxia, tremor, dysarthria, external ophthalmoplegia, facial palsy, rigidity, and dementia**

   2. Radiologic imaging studies (and gross inspection of autopsy brain) reveal marked **atrophy of basis pontis, inferior olivary nucleus, and cerebellum**

   3. Genetic aspects

      a. Various hereditary patterns have been found in different families, suggesting category of disorders rather than single disease process

      b. Autosomal dominant spinocerebellar ataxia type 1 — one form related to abnormality on short arm of chromosome 6 associated with trinucleotide (CAG) repeat sequence

   C. **Ataxia-telangiectasia (Louis-Bar syndrome)**

   1. Progressive **gait ataxia** beginning in infancy with subsequent development of **limb ataxia, choreoathetosis**, dysarthria, **nystagmus, areflexia**, and development of **telangiectasia** (blood vessel dilation) of conjunctiva, face, and neck

Chapter 6

2. Associated with **thymic and lymphoid hypoplasia**, **deficiency of serum IgA**, recurrent **sinopulmonary infections**, and development of **lymphoreticular malignancies**, generally resulting in death during childhood

3. **Autosomal recessive inheritance**; genetic defect involves inability to repair DNA damage produced by radiation or intracellular metabolites

# Chapter 7 — COMA AND IMPAIRED CONSCIOUSNESS

I. **Impaired consciousness** — various neurologic conditions result in impaired consciousness, which can often progress to death; thus, these are medical emergencies necessitating immediate evaluation along with treatment of reversible conditions

II. Terms related to coma and altered consciousness

   A. **Consciousness — state of awareness and arousal**

   1. **Awareness** — **cerebral cortical** ability to meaningfully **interpret sensory input** allowing proper understanding of self and environment

   2. **Arousal** — responsiveness to environment mediated by **brain stem network** extending from medulla to thalamus: **ascending reticular activating system** projects diffusely to cerebral cortex and acts as "on-off switch" for cortical awareness (for example, in sleep-wake cycle)

   B. **Level of consciousness** — ranges from full alertness to total nervous system unresponsiveness; **altered consciousness** results from either **bilateral cerebral hemispheric disturbance** or **brain stem disturbance**; altered consciousness does *not* occur after only unilateral cerebral hemispheric injury (unless opposite hemisphere or brain stem are also affected, such as by edema); levels of consciousness are divided (for ease of clinical assessment) into several discrete categories:

   1. **Alert, awake, and oriented** — normal state of consciousness

   2. **Confusion (confusional state)** — responsive to stimuli, but disoriented with respect to time, person, and place

      a. **Inattention** — inability to select among sensory stimuli resulting in almost random responses

      b. **Delirium** — acute confusional state; sometimes also associated with drowsiness

   3. **Drowsiness** — **inclination to sleep; easily aroused** and able to respond to most stimuli both verbally and with motor defenses (fending off stimulus)

   4. **Stupor** — **little spontaneous physical or mental activity; reduced responsiveness** to environmental stimuli; generally unresponsive to verbal stimuli, and only partially

Chapter 7

arousable to vigorous (usually painful) stimulation; minimal or no verbal response, but able to respond with motor defenses

5. **Semicoma (light coma)** — **appears to be sleeping**, but **generally unresponsive** to all but most vigorous stimulation; **only primitive reflexes** and abnormal body posturing occur

6. **Coma (deep coma)** — **lack of responses to even most painful stimuli**; absence of even primitive reflexes

C. **Psychogenic unresponsiveness** — **psychiatric conversion symptoms** superficially suggesting coma, but with physiologic evidence of wakefulness and no disturbance of brain function

D. **Persistent vegetative state (coma vigil)**

1. Characterized by **spontaneous eye opening** (and sometimes in response to sound), spontaneous eye movement (as though following objects in environment), and electroencephalographic (EEG) pattern of nearly **normal sleep-wake cycle**, but **no evidence of responsiveness to or awareness of environment**; "arousal without awareness"

2. Results from severe diffuse cerebral cortical damage

E. **Locked-in syndrome**

1. Characterized by **awareness and wakefulness**, but **inability to respond or communicate except with eye movement**

2. Results from large brain stem lesion beginning just caudal to oculomotor nerve (cranial nerve III) and destroying basis pontis (thereby interrupting corticobulbar and corticospinal pathways), but sparing ascending reticular activating system and sensory pathways in pontine tegmentum

3. Alexander Dumas (*The Count of Monte Cristo*) described this condition in his character M. Noirtier de Villefort as "a mind...clogged by a body rendered utterly incapable of obeying its impulses...a corpse with living eyes."

| GLASGOW COMA SCALE | |
|---|---|
| **Eye opening** | |
| Never | 1 |
| To pain | 2 |
| To verbal stimuli | 3 |
| Spontaneously | 4 |
| **Best Verbal Response** | |
| No response | 1 |
| Incomprehensible sounds | 2 |
| Inappropriate words | 3 |
| Disoriented and converses | 4 |
| Oriented and converses | 5 |
| **Best Motor Response** | |
| No response | 1 |
| Extension (decerebrate rigidity) | 2 |
| Flexion abnormal (decorticate rigidity) | 3 |
| Flexion withdrawal | 4 |
| Localizes pain | 5 |
| Obeys | 6 |
| TOTAL SCORE (range) | 3–15 |
| Sum of highest value in each category is coma score: Full mental capacity = 15  Highest level of coma = 8  Brain death = 3 | |

# COMA AND IMPAIRED CONSCIOUSNESS

III. Evaluation and treatment of coma or altered consciousness

A. Immediate stabilization of airway, ventilation, and circulation

B. Determination of level of consciousness — **Glasgow Coma Scale** provides objective score useful for following level of consciousness

C. Evaluation of head and eye position — unilateral hemispheric damage results in **conjugate eye and head deviation** (turning toward side of hemispheric lesion and away from side of hemiparesis); unilateral pontine damage results in conjugate eye and head deviation (toward side of hemiparesis and away from side of lesion); "**eyes look at cerebral lesion and away from pontine lesion**"

D. Evaluation of pupillary size and reaction to light

1. Reactive pupils — indicates intact midbrain function; reactive pupils associated with absent extraocular movements suggests metabolic disturbances such as hypoglycemia or barbiturate intoxication

2. **Ipsilateral pupillary dilation** (and nonreactivity to light) — indicates **compression of oculomotor nerve** (cranial nerve III) by **transtentorial uncal herniation**

3. **Midposition nonreactive pupils** — **midbrain damage** suggested by pupils of 3-5 mm diameter, symmetric, and unreactive

4. **Bilaterally fixed and dilated pupils** — severe **structural midbrain damage** or metabolic disturbance due to glutethimide, scopolamine, or atropine poisoning

5. **Pinpoint pupils** — **narcotic overdose**; following treatment with narcotic antagonist such as naloxone (Narcan), pupils should dilate and become reactive

E. Evaluation of breathing pattern

1. **Cheyne-Stokes respiration** — **periodic breathing** with smooth transition from apnea to hyperpnea to apnea; usually associated with bilateral lesions deep in cerebral hemispheres or with metabolic disorders

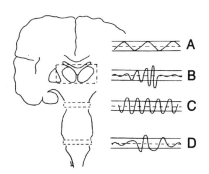

Breathing patterns noted with coma: (A) normal; (B) Cheyne-Stokes respiration; (C) central neurogenic hyperventilation; (D) ataxic breathing.

2. **Central neurogenic hyperventilation** — sustained, regular, **rapid, deep hyperpnea**; found with midbrain dysfunction

3. **Ataxic breathing — irregular gasping** breathing; found with medullary damage

F. Evaluation of extraocular movements — in unresponsive patients, testing of **vestibulo-ocular reflexes** assesses intactness of brain stem from medullary vestibular nuclei to midbrain oculomotor nuclei

1. **Doll's eye maneuver (oculocephalic reflex)** — elicited by holding patient's eyelids open and passively rotating head rapidly to each side [must not be performed if there is possible cervical spine injury]

    a. Awake patients have suppression of oculocephalic reflex

    b. In unconscious patient with intact reflex bilaterally, **eyes move in direction opposite to head** (as though visually-fixed on object); this indicates intact brain stem between oculomotor and vestibular nerves and suggests that decreased level of consciousness is not due to structural brain stem lesion

With doll's eye maneuver in an unconscious patient, eyes move in direction opposite to head.

    c. Unilateral reflex (eye movement in only one direction) indicates destructive brain stem lesion

    d. Absent response (no eye movement) indicates either destructive brain stem lesion or barbiturate poisoning

2. **Caloric testing (ice water calorics)** — when doll's eye maneuver produces equivocal information or when patient's neck should not be rotated (as in cervical spine injury), caloric testing can be performed

    a. External auditory canal must not be blocked by blood or wax and tympanic membrane must be intact; 50 cc of ice water is instilled into external auditory canal of one ear, while head (or head of bed) is elevated 30 degrees; procedure can be performed in opposite ear after five minute interval

In unconscious patient, instillation of ice water in ear results in tonic deviation of eyes toward stimulated ear.

    b. Awake patient develops nystagmus with slow deviation of both eyes toward stimulated ear (brain stem response) and fast corrective movement toward opposite ear (frontal lobe response)

    c. **Unconscious patient** with normal response has only **slow deviation of both eyes toward each stimulated ear**, indicating that brain stem between cranial nerve III (oculomotor nerve) and cranial nerve VIII (vestibuloacoustic nerve) is intact and indicating that decreased level of consciousness is not due to structural brain stem lesion

    d. Unilateral response (movement of only one eye or response to stimulation in only one ear) indicates structural brain stem lesion

    e. Absent response bilaterally indicates either structural brain stem lesion or barbiturate coma

G. Examination of limbs for paralysis, asymmetry of tendon reflexes, or Babinski reflexes (extensor plantar responses), or reflex posturing (movements); herniation produces progression from decorticate to decerebrate posturing

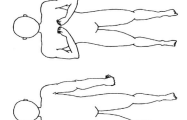

Limb positions in decerebrate and decorticate posturing.

    1. **Decorticate posturing — flexion and adduction of arms and extension of legs**; suggests lesion **above midbrain involving both cerebral hemispheres** (usually deep midline lesion)

    2. **Decerebrate posturing** — extension, adduction, and internal rotation of arms and extension of legs; suggests lesion involving upper brain stem **caudal to midbrain red nucleus**

H. Evaluation for possible metabolic abnormalities or toxins that could cause or contribute to unconsciousness

    1. Blood and urine samples for chemical analysis

    2. Administration of intravenous glucose (for hypoglycemia), thiamine (for Wernicke's encephalopathy), and naloxone (for possible narcotic overdose)

I. Radiologic imaging studies to investigate structural causes for unconsciousness

J. Evaluation and monitoring of intracranial pressure (including funduscopic examination for evidence of papilledema) and treatment of edema and increased intracranial pressure with corticosteroids (dexamethasone), hyperventilation, intravenous hyperosmolar agents (mannitol), ventricular drainage, and barbiturates

IV. **Herniation syndromes**

    A. Rigid membranes (falx cerebri and tentorium cerebelli) separate cranial contents into three major compartments (left hemisphere, right hemisphere, and posterior fossa); displacement (from mass effect of blood, edema, or tumor) of brain tissue from one compartment to another or through foramen magnum is termed herniation

    B. Herniation is ominous since brain structures are distorted and compressed

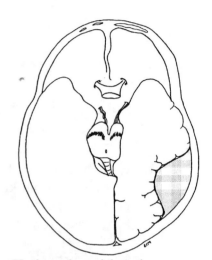

Horizontal cranial section showing right subdural mass producing transtentorial herniation with resultant oculomotor nerve and midbrain compression.

        1. **Compression of blood vessels** causes additional damage from ischemia; brain death results when carotid and vertebral artery flow to cranial cavity ceases due to vascular compression

        2. Distortion and **tearing of penetrating blood vessels** originating from **basilar artery** results in **midbrain and pontine hemorrhages (Duret hemorrhages)**

        3. **Compression of cranial nerves** results in loss of function

        4. **Medullary compression damages respiratory and cardiovascular control** centers with consequent apnea and cardiovascular collapse

    C. **Transtentorial uncal herniation**

        1. Produced by **unilateral cerebral hemispheric mass** which pushes **inferomedial portion of temporal lobe (temporal lobe uncus) through tentorial opening** (tentorial incisura) and against midbrain

        2. **Oculomotor nerve (cranial nerve III) is compressed** causing initial **ipsilateral pupillary dilation**, followed later by loss of extraocular movements (eye will be **deviated down and out** due to residual function of external rectus muscle innervated by abducens nerve and superior oblique muscle innervated by trochlear nerve)

        3. Compression and distortion of **ipsilateral cerebral peduncle** results in **contralateral hemiparesis**

        4. Midbrain distortion interferes with **ascending reticular activating system** resulting in loss of consciousness

5. Compression of posterior cerebral artery results in **hemorrhagic infarction of occipital and posterior temporal lobes**

6. As herniation continues, opposite sharp edge of tentorium cuts into contralateral cerebral peduncle resulting in focal necrosis (Kernohan's notch) and **ipsilateral hemiparesis**; further brain stem distortion abolishes brain stem reflexes and is associated with **brain stem hemorrhages (Duret hemorrhages)**; medullary dysfunction results in cardiovascular collapse and death

D. **Central rostral-caudal herniation**

1. **Bilateral cerebral hemispheric masses** or edema or deep central cerebral or thalamic masses result in **central cerebral structures pushing downward** on midbrain

2. **Midbrain distortion** results initially in **fixed midposition pupils**, loss of consciousness, and **bilateral decorticate posturing**

3. As herniation progresses, pupils dilate and ophthalmoplegia occurs; further brain stem distortion abolishes brain stem reflexes and is associated with brain stem hemorrhages (Duret hemorrhages); medullary dysfunction results in cardiovascular collapse and death

E. **Subfalcial (cingulate) herniation**

1. **Unilateral frontal lobe mass** lesion pushes **anterior cingulate gyrus under falx cerebri**

2. Clinically symptomatic only if **anterior cerebral artery compression** occurs, resulting in hemorrhagic infarction in medial superior frontal lobe with clinical signs of leg weakness

F. **Cerebellar tonsillar (foramen magnum) herniation**

1. Expanding **mass lesion in or around cerebellum** results in compression of pons and medulla

2. Initial presentation of **vomiting, dizziness, ataxia, marked "malignant" hypertension, drowsiness, leg weakness, and spasticity**

3. **Cerebellar tonsils herniate** through foramen magnum producing cerebellar tonsillar necrosis, medullary compression, and distortion; signs include tonic deviation of eyes away from side of lesion, ataxic (irregular) breathing, coma, and death

4. Progresses very rapidly because there is much less space in posterior fossa than in supratentorial compartment; thus, in treatable conditions (such as cerebellar hemorrhage) immediate neurosurgical decompression is required to prevent death

V. **Brain death — irreversible cessation of all functions of entire brain**, including brain stem (but not spinal cord)

   A. Criteria

   1. **Established coma-producing cerebral lesion** associated with **irreversible** widespread damage

   2. **Absence of toxins or metabolic abnormalities** that could produce coma; barbiturates, tranquilizers, and neuromuscular junction blockers must particularly be excluded; renal and hepatic failure preclude diagnosis of brain death

   3. Body **temperature greater than 90°F (32°C)**

   4. **Complete unresponsiveness** — no evidence of spontaneous or reflex activity from cerebrum or brain stem; no response to noxious or other stimuli (including changes in heart rate); unreactive pupils (dilated or midposition); absent corneal, vestibulo-ocular, gag, and all other brain stem reflexes

   5. **Apnea** — must be confirmed by demonstrating lack of breathing despite hypercapnia (arterial carbon dioxide levels greater than 60 mm Hg)

   6. Spinal reflexes can be present, but **no reflexes involving structures above level of cervical spinal cord**

   7. **Criteria must be met for at least 6 hours** (with confirmatory test) or 12-24 hours without such test

   B. Confirmatory tests

   1. **Isoelectric electroencephalogram** — absent EEG activity must be demonstrated by testing performed according to rigorous standards established by American Electroencephalographic Society; presence of EEG activity precludes brain death

   2. Brain stem auditory evoked potentials — absent potentials provides evidence of absent brain stem function

   3. **Absence of cerebral blood flow** — demonstrable with cerebral angiography or bolus radionuclide angiography; funduscopic evidence of sludging of blood in retinal veins bilaterally indicates absent cerebral blood flow

   C. Brain death is declared by physician and documented in records, before any therapy is stopped

# Chapter 8                           CENTRAL NERVOUS SYSTEM TRAUMA

I. **Head trauma**

    A. Brain is soft and readily disrupted; protection by rigid skull and cerebrospinal fluid jacket also makes possible generalized agitation following heavy blows and provides no room for brain swelling.

    B. Terms for **craniocerebral trauma**:

        1. **Missile injury** — damage produced by moving object striking cranium; most often refers to bullet injury

        2. **Penetrating (open) injury** — disruption (penetration) of cranial vault with opening through skin and cranial bones exposing damaged brain; most often associated with missile injury

        3. **Closed (non-penetrating) injury** — damage to brain without disruption or skin over cranial vault; most often results from blunt trauma

        4. **Acceleration/deceleration injury** — damage produced by movement of brain within confines of cranial vault; brain injury results from tearing during violent movement or from impact of striking interior of skull or dural folds

        5. **Compressive injury** — damage resulting from compression of cranial vault with resultant fracture of bones and injury to underlying brain

II. **Missile injuries** — produced by objects that fall or are propelled through air, striking cranium

    A. Damage related to amount of energy transmitted to brain

        1. Energy of missile is proportional to mass and to square of velocity: thus, faster lighter missiles can do more damage than slower heavier missiles

        2. For example: missile weighing 2.5 grams [approximate weight of slug from .22 caliber rifle bullet] traveling at 1100 ft/sec [muzzle velocity of .22 rifle] has energy of about 150 joules; in contrast, approximately same size missile traveling at 3200 ft/sec [muzzle velocity of U.S. military M-16 rifle firing slug nearly same size as that of .22 caliber rifle] has energy of about 1200 joules

    B. Extent of injury depends on missile energy, shape, and direction of impact

1. **Tangential injuries** — scalp lacerations, depressed skull fractures, and laceration of underlying brain

2. **Penetrating injuries** — skin, hair, bone, and missile fragments driven into brain; low velocity missiles (less than 1100 ft/sec) may **ricochet on inner table of skull**, thereby widening area of damage

3. **Through-and-through injuries** — usually associated with high velocity missiles; in addition to driving skin, hair, and bone fragments into brain, **shock waves** stretch, shear, or rupture blood vessels, nerves, and bone at considerable distance from missile entry or exit site

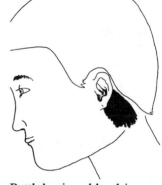

Battle's sign: blood in soft tissue overlying mastoid; often associated with blood in external auditory meatus.

III. **Closed head injury**

A. **Scalp injuries** — abrasions or lacerations require local cleansing and suturing to close any skin openings; scalp lesions may be only reliable evidence of underlying injuries to skull and/or brain

B. **Skull fracture** — suggests considerable force of impact; most fractures are linear (fissure) fractures

1. **Depressed fracture** — bone **displaced** toward brain

2. **Compound fracture** — **laceration of scalp** associated with skull fracture

Raccoon eyes: blood in periorbital soft tissues.

3. **Basilar skull fracture** — fracture passing through base of skull which may damage cranial nerves or blood vessels normally passing through skull foramina

   a. **Battle's sign** — fractures through **petrous bone** (ear) producing **subcutaneous blood over mastoid**; usually associated with tympanic membrane rupture and blood drainage and/or cerebrospinal fluid leakage into middle ear (**CSF otorrhea**) or out external auditory canal; also usually associated with damage to vestibuloacoustic nerve (cranial nerve VIII)

   b. **"Raccoon eyes"** — fractures through **anterior cranial fossa** result in accumulation of **subcutaneous blood around eyes**; often associated with cerebrospinal fluid leakage into sinuses (**CSF rhinorrhea**); damage to olfactory nerve (cranial nerve I) common, and damage to optic nerve (cranial nerve II) and oculomotor (cranial nerve III) can occur

c. **Carotid-cavernous fistula** — fractures associated with **tearing** of **carotid artery within cavernous sinus**; results in massive **shunting** of blood producing venous distention; presents as **painful, pulsating exophthalmos**

d. Pneumocele and CSF leakage — fractures through inner wall of nasal sinuses or mastoid air cells permit air to enter cranial cavity (visible on radiologic imaging studies) along with bacteria (which can produce meningitis and/or brain abscess); persistent CSF leakage requires surgical repair to prevent recurrent infection

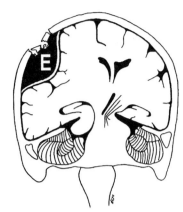

Epidural hematoma.

C. **Epidural hematoma** — **arterial** hemorrhage from **temporal bone fracture** which **tears middle meningeal artery**

1. Clinical presentation — initial unconsciousness (due to cerebral concussion); usually followed by arousal (**"lucid interval"**), although up to 50% of patients may not arouse due to severe head injury; then followed by progressive **headache** and **drowsiness**, **hemiparesis**, **dilating pupil** on side of hemorrhage (indicating **transtentorial uncal herniation** with compression of oculomotor nerve), and slowing of pulse and respirations with increase in blood pressure (**Cushing's reflex** indicating brainstem distortion with medullary dysfunction from herniation)

2. Arterial damage results in brisk bleeding, creating rapidly expanding mass **between dura and cranial bone** (thus, *epi*dural)

3. Death ensues unless bleeding is controlled and mass removed; requires immediate neurosurgical evacuation of clot, which usually produces complete recovery; survivors of delayed intervention usually have permanent brain damage

D. **Subdural hematoma** — **venous** hemorrhage from damage to venous sinuses or to veins communicating between cerebral cortex and venous sinuses (**"bridging veins"**)

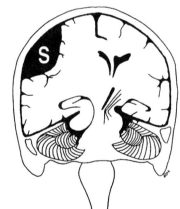

Subdural hematoma.

1. Commoner than epidural hematoma, since only minimal trauma is necessary to tear veins (often without fracture; sometimes trauma is only incidental without external evidence); venous sinus tears usually result from skull fractures

2. Consists of accumulation of blood **between arachnoid and dura** (thus, *sub*dural); most often located over lateral (convexity) surface of frontal and parietal lobes

3. Pathophysiology

   a. Small subdural hematomas can be clinically insignificant and slowly absorbed; larger hematomas can undergo **gradual enlargement** over several weeks (even though active bleeding has stopped) resulting in brain compression; although enlargement is gradual, surgically evacuation is necessary to prevent brain herniation and death

   b. Response of dura to subdural hemorrhage is organization with formation of delicate fibrous **neomembrane**; enlargement probably relates to several factors: abnormal permeability of vessels in neomembrane, rupture of delicate vessels in neomembrane with **recurrent bleeding** ("rebleeding"), and increased osmotic pressure from blood breakdown products

4. **Acute subdural hematoma** — damage to **venous sinus** results in rapid accumulation of blood with clinical presentation mimicking epidural hematoma and necessitating prompt neurosurgical evacuation and control of bleeding

5. **Chronic subdural hematoma** — slower accumulation of blood due to tearing one or more **bridging veins**; bleeding ceases spontaneously (probably after few hours) due to tamponade

   a. Slowly progressive clinical symptoms of headache, confusion, hemiparesis, apathy, lethargy, and ultimately coma

   b. Radiologic imaging studies confirm diagnosis; MRI preferable, since on CT scans initial hyperdensity of hematoma (denser appearance than adjacent brain) changes to isodensity (similar appearance to adjacent brain) making detection difficult

   c. Treatment involves neurosurgical drainage

   d. Common complication of falls in infancy or of child abuse; clinical presentation of increased intracranial pressure with enlarging head (increased head circumference), bulging fontanelle, vomiting, irritability, and lethargy; often associated with seizures; in child abuse, fractures of skull and other bones may be evident; associated retinal hemorrhages without evidence of external trauma suggests **"shaken baby syndrome"**

E. **Brain injury**

1. **Concussion** — following head injury, there is **temporary** impairment of cerebral neuronal function without evidence of structural damage (example is "10-second knockout" in boxing)

# CENTRAL NERVOUS SYSTEM TRAUMA

a. Can be only consequence of head injury or can accompany more severe injuries (such as skull fractures and epidural hematoma)

b. Following recovery, some patients experience "postconcussion syndrome" of headache, dizziness, and personality changes

2. **Contusion and laceration** — impairment of brain function due to **bruising and tearing of brain tissue**, with resultant necrosis, hemorrhage, and edema

   a. Focal contusions occur under site of impact (**"coup" lesion**), but also at opposite side (**"contrecoup" lesion**) particularly on **undersurface of frontal lobe or anterior part of temporal lobe** from bouncing against bony irregularities of orbital roof or lesser wing of sphenoid

   b. Small subpial, cortical, and superficial white matter vessels rupture (characteristically over tips of gyri) with resultant wedge-shaped hemorrhages (apex directed into white matter) often visible on radiologic imaging studies (particularly MRI)

   c. **Plaque jaune** — brownish-yellow, glial-collagenous **scar** formed at sites of superficial **cortical contusions**

   d. Large intracerebral hemorrhages or hemorrhages into deep nuclei (basal ganglia or thalamus) may also occur

   e. **Diffuse axonal injury** — disruption of **axons in long tracts** (such as those between cerebral cortex and brain stem), attributable to shear and tensile strains from acceleration/deceleration; results in prolonged **coma** or death

IV. Treatment of head injury — initial management depends on whether patient is alert or comatose; alert patient may need only careful neurologic examination and radiologic imaging study, but comatose patient needs urgent evaluation and treatment

   A. Stabilization of airway, ventilation, and circulation must be accomplished immediately

   B. Determination of level of consciousness — **Glasgow Coma Scale** good predictor: 87% with score of more than 11 in first 24 hours had good recovery, while 87% with scores of less than 5 remained in vegetative state or died

   C. Evaluation of **pupillary size** and **reaction to light** — **transtentorial uncal herniation** indicated by **ipsilateral pupillary dilation** (and nonreactivity to light) followed subsequently by similar changes in contralateral pupil

   D. Examination of limbs for paralysis, asymmetry of tendon reflexes, or Babinski reflexes (extensor plantar responses) indicating corticospinal tract damage

Chapter 8

- E. Radiologic evaluation for possible spinal injury

- F. Seizure prevention and control with intravenous anticonvulsant phenytoin

- G. Assessment for other possible injuries — shock suggests associated injury with hemorrhage into chest, abdomen, pelvis, retroperitoneum, or thigh

- H. Consideration of explanations for unconsciousness other than head injury (for example, shock, alcohol or drug intoxication, adrenal insufficiency, stroke, diabetic acidosis, insulin-induced hypoglycemia, or other metabolic disturbances)

- I. Assessment of potential neurosurgically-treatable conditions such as epidural or subdural hematoma

- J. Monitoring intracranial pressure and treatment of increased pressure with corticosteroids (dexamethasone), hyperventilation, intravenous hyperosmolar agents (mannitol), ventricular drainage, and barbiturates

V. **Delayed complications** of head injury

- A. **Post-traumatic epilepsy**

    1. Result of contusion-laceration of cerebral cortex, particularly from contrecoup injuries to **orbital frontal**, **inferior temporal**, or **inferior frontal cortex**

    2. Seizures following head injury can be classified as:

        a. **Immediate (at time of injury** or within minutes of injury) — usually not associated with later development of epilepsy

        b. **Early (within first week** following injury) — particularly common in children and usually associated with later epilepsy

        c. **Late (after first week** following injury) — referred to as post-traumatic epilepsy; onset usually by 2 years following injury

- B. Occurs in about 5% of all patients with closed head injuries and up to 50% with depressed skull fracture or penetrating injuries

- C. **Postconcussion syndrome — headache, dizziness, and personality changes** following concussive head injury

    1. Chronic headache — constant or throbbing; often resembles migraine

# CENTRAL NERVOUS SYSTEM TRAUMA

2. Dizziness — can be true vertigo related to vestibular nerve damage

3. Amnesia, poor concentration, learning difficulties, emotional lability, irritability, aggressiveness, and hyperactivity

4. Does not correlate with duration of unconsciousness, but pre-existing (prior to head trauma) psychiatric problems (such as depression) and pending litigation appear to be predisposing factors

D. **Post-traumatic hydrocephalus** — blood in subarachnoid space induces scarring (**arachnoidal fibrosis**) which obstructs cerebrospinal fluid pathways resulting in hydrocephalus; diagnosis based on progressive headaches, confusion, lethargy, and radiologic imaging studies showing **large ventricles**; neurosurgical insertion of **ventricular shunt** usually produces prompt improvement of symptoms

E. **Post-traumatic dementia** (**dementia pugilistica**, "punch drunk syndrome")

1. Repeated cerebral injuries (as occurs in boxers) results in syndrome of dementia, often developing many years after last injury

2. Characterized by **memory disturbance**, **confusion**, and **dysarthria**; slow movements and shuffling, wide-based gait with other parkinsonian features can also be evident

3. Radiologic imaging studies demonstrate cerebral cortical atrophy, ventricular dilation, and cavum septi pellucidi

4. Neuropathologic studies have shown **thinning of corpus callosum**, enlargement of lateral ventricles, **cavum septi pellucidi**, **loss of pigmented neurons** of substantia nigra and locus ceruleus, and presence of **neurofibrillary tangles** in cerebral cortical and brain stem neurons (**without any neuritic plaques**)

VI. **Spinal cord injury** — compression, concussion, contusion-laceration, or transection

A. Most common sites of traumatic lesions are **cervical cord** (particularly lower cervical, C5-T1) and **thoracolumbar junction** (T11, T12, L1)

B. **Spinal cord concussion**

1. Similar to brain concussion and characterized by temporary impairment of spinal cord function without evidence of vertebral fracture or dislocation

2. **Initial flaccid paralysis and sensory loss** followed by **complete recovery** with no apparent structural damage to cord

Chapter 8

3. Fracture-dislocation or compressive lesions must be ruled out

C. **Spinal cord contusion-laceration**

1. Similar to brain contusion-laceration with impairment of spinal cord function due to **bruising and tearing** of tissue and resultant necrosis, hemorrhage, and edema

2. Early changes relate to small hemorrhages, necrosis, and edema with subsequent resolution to scarring and formation of cysts or **syringomyelic cavities**

3. Clinical symptomatology relates to degree and level of injury, varying from complete flaccid paralysis and sensory loss to focal deficits only

   a. Functional **cord transection**

      (1) **Spinal shock — bilateral flaccid paralysis and sensory loss** below level of lesion, areflexia, and bladder, bowel, and anal sphincter dysfunction, usually lasting several weeks

      (2) Slow recovery to increased reflex activity below level of lesion; **mass action reflex** consisting of leg triple flexion (ankle extension, knee flexion, and hip flexion), abdominal muscle contraction, profuse sweating, piloerection, and automatic urination (and occasionally defecation) in response to stimulation below level of lesion

      (3) **Autonomic disturbances** include profuse sweating, orthostatic hypotension and/or paroxysmal hypertension, bladder dysfunction (initial atonic bladder, followed later by automatic reflex or spastic bladder which empties with small volume), and bowel dysfunction (initial distention and lack of peristalsis, followed later by automatic reflex defecation)

      (4) Some recovery of sensation and/or development of pain (described as burning) with incomplete lesions

   b. **Anterior cord syndrome — paralysis** and **loss of pain** and **temperature** sensation but **preserved proprioception** (vibration and position sense) below level of lesion; due to compression or damage to anterior part of cord as from intervertebral disc fragment

   c. **Central cervical cord syndrome — greater weakness in arms than in legs**, patchy sensory loss below level of lesion, and urinary retention; often results from **neck hyperextension**

d. **Brown-Séquard syndrome (cord hemisection)** — **ipsilateral spastic weakness, ipsilateral loss of proprioception** (vibration and position sense), and **contralateral loss of pain and temperature** sensation

D. Treatment of spinal cord injuries involves establishing diagnosis with neurologic examination and radiologic imaging studies, while maintaining stability (preventing motion) in spinal column and subsequently repairing vertebral column damage; use of traction may be particularly valuable in cervical injuries, but surgical fixation is usually required to assure stability

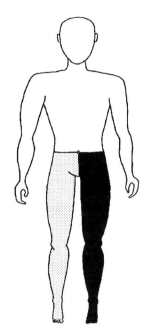

Brown-Séquard syndrome from left T10 cord hemisection with ipsilateral (left) spastic weakness and loss of proprioception and contralateral (right) loss of pain and temperature sense.

1. **Atlanto-occipital dislocation** — often immediately fatal due to vertebral artery tearing or medullary compression

2. **Jefferson fracture** — blow to vertex of head resulting in **split posterior arch of C1 (atlas)**; if not immediately fatal due to cord concussion, can have no associated neurologic signs since canal is widened by fracture

3. **Odontoid injuries** — **separation of odontoid** may be fatal due to medullary compression, either immediately or following subsequent neck movement

4. **Hangman's fracture** — **disruption of arch of C2** along with **C2-C3 dislocation**; usually results from hyperextension

5. **Locked facets** — facet dislocation; unilaterally associated with root injury; bilaterally associated with cord injury

6. Dislocations through disc space or vertebral body — marked misalignment with severe cord injury

7. **Wedge fractures** — common in **thoracic** vertebral injuries

8. **Vertebral body bursting** — common with **thoracolumbar** vertebral injuries; results in **free floating bone fragments** in spinal canal pressing on spinal cord and roots

# Chapter 9  PERIPHERAL NERVOUS SYSTEM DISORDERS

I. Normal peripheral nervous system

   A. Division of peripheral from central nervous system is artificial, since they are really continuous: cell bodies of anterior horn neurons are located in spinal cord (central nervous system), while their axons are located in ventral roots and peripheral nerves (peripheral nervous system); dorsal root ganglion neuron cell bodies and their peripheral axons are considered part of peripheral nervous system, while their "central" axons (in spinal cord dorsal columns) are part of central nervous system; diseases often do not respect notions of "boundaries" between central nervous system and peripheral nervous system

   B. Nerve axons can be either myelinated or unmyelinated

      1. Single **oligodendrocytes** produce myelin internode segments for multiple **central nervous system axons**

      2. Each **peripheral nervous system myelin segment (internodal segment)** is formed by single **Schwann cell** cytoplasm; unmyelinated axons are also encased by cytoplasm of Schwann cells which do not form myelin; bundles of myelinated and unmyelinated fibers form **fascicles** enveloped by specialized connective tissue sheath (**perineurium**) which produces **blood-nerve barrier** (analogous to blood-brain barrier)

   C. Muscle is composed of multiple muscle fibers enclosed by fascial connective tissue sheath; muscle fibers are multinucleate cells containing thousands of cytoplasmic (sarcoplasmic) **myofibrils** composed of **actin** and **myosin** (contractile apparatus); each muscle fiber has single **acetylcholine receptor-rich endplate** opposite motor nerve terminal; muscles are mosaics of two distinctive fiber types

      1. **Type 1 muscle fibers** — slow twitch, red, aerobic, oxidative, and fatigue-resistant

      2. **Type 2 muscle fibers** — fast twitch, white, anaerobic, glycolytic, rapidly-fatiguing

II. Clinical terminology

   A. **Mononeuropathy** — a local process (such as trauma, entrapment, vascular disease, or infection) involving **dysfunction of single nerve**; **pain** is common

   B. **Mononeuropathy multiplex** — multifocal (**multiple nerve**) dysfunction; generalized, but **asymmetrical** process involving two or more nerves ("many single nerves"); usually, results from **compromise of vasonervorum** with consequent **nerve infarction** (such as in polyarteritis

## PERIPHERAL NERVOUS SYSTEM DISORDERS

nodosum, diabetes mellitus, systemic lupus erythematosus, Wegner's granulomatosis, or rheumatoid arthritis); presents as abrupt onset of **pain** along with sensory symptoms and/or motor signs

C. **Polyneuropathy** — generalized **symmetrical** multiple nerve dysfunction; usually **distal greater than proximal** with longer nerves affected most (for example, in polyneuropathy associated with uremia or alcoholism)

D. **Radiculopathy** — dysfunction of spinal **nerve root**; can be isolated finding (such as in trauma or intervertebral disc protrusion) or can be part of polyneuropathy or mononeuritis multiplex.

E. **Cranial neuropathy** — dysfunction of one or more **cranial nerves** can be isolated finding involving single cranial nerve (such as diabetic ophthalmoplegia due to oculomotor nerve infarction) or part of generalized polyneuropathy

F. **Ganglionopathy** — dysfunction of **nerve ganglion**, usually as part of generalized process

G. **Autonomic neuropathy** — dysfunction of **autonomic neurons** (with consequent anhidrosis, orthostatic hypotension, pupillary reflex paralysis, loss of lacrimation and salivation, impotence, and bowel and bladder dysfunction); most commonly occurs as a part of a **generalized polyneuropathy** (such as in diabetes mellitus), but also as an autosomal recessive disease (**Riley-Day syndrome**) or as part of a parkinsonian neurodegenerative disorder (**Shy-Drager syndrome**)

H. **Plexopathy** — dysfunction of part or all of **brachial or lumbar plexus**; can be idiopathic or result from trauma, local compression, tumor, infection or delayed effect of radiotherapy

I. **Amyotrophy** — term used to describe **muscle atrophy** secondary to damage to innervating nerve axons; usually applied to diabetic amyotrophy, neuralgic amyotrophy, and amyotrophic lateral sclerosis

J. **Myopathy** — **muscle weakness** from abnormality of muscle fibers (nerve axons are presumed to be normal)

K. **Myositis** — muscle destruction (and accompanying clinical weakness) caused by **inflammation** (may be infectious or autoimmune)

L. **Motor neuropathy** — dysfunction of **motor axons** either from direct damage to axons or due to death of anterior horn cells

M. **Sensory neuropathy** — dysfunction of **sensory nerve fibers**

N. **Sensorimotor neuropathy** — most common form of neuropathy in which both **sensory and motor axons** are dysfunctional

Chapter 9

III. Clinical symptomatology

    A. **Weakness** — must be objectively confirmed since term frequently used by patient to describe other kinds of subjective experience

    B. **Hypesthesia — diminished sensibility** to stimulation (may involve one or more sensory modalities)

    C. **Paresthesia** — spontaneous **aberrant** (commonly "pins and needles") sensory experience

    D. **Dysesthesia — disagreeable** sensory experience

    E. **Hyperesthesia — increased** intensity of subjective sensory experience following stimulation; usually associated with objectively diminished sensibility

IV. Signs

    A. **Weakness** — usually **distal and symmetrical in polyneuropathy**; usually **proximal in myopathy**

    B. **Muscle atrophy — loss of muscle bulk** resulting from loss of neural input; associated with **reduced tone** and flaccidity.

    C. **Sensory deficit** — varies with modality or with somatotopic distribution

    D. **Areflexia — loss of reflexes**; occurs later in pure motor neuropathy than in mixed (sensorimotor) or pure sensory neuropathy (contrasts with hyperreflexia associated with damage to "upper motor neuron" or corticospinal tract); areflexia usually occurs late in myopathy when weakness precludes limb movement.

V. Electrodiagnostic tests

    A. **Nerve conduction velocity (NCV)** — recording of speed of propagation of electrical impulse by **largest myelinated axons** in nerve; **F-wave latency** evaluates conduction velocity in **proximal** nerve and nerve root; **H-reflex** evaluates **reflex arc** in lower extremity

        1. Loss of nerve (axon) continuity leads to failure of impulse conduction

        2. Demyelination of otherwise intact axons leads to slowing of conduction.

    B. **Electromyography (EMG)** — recording of electrical activity of muscle fibers

# PERIPHERAL NERVOUS SYSTEM DISORDERS

1. Motor unit potentials of increased amplitude suggest reinnervation of denervated muscle fibers by axonal sprouting from adjacent intact nerves, producing enlarged motor units with increased numbers of muscle fibers per anterior horn cell

2. **Fasciculations** are grossly **visible muscle twitches** resulting from uncoordinated firing of muscle fibers in a single motor unit due to excessively irritable motor nerve

3. **Fibrillations** are twitchings of **single denervated muscle fibers** (visible by surgical exposure of muscle belly)

VI. Pathophysiology of nerve disease

   A. **Axonal degeneration** (regressive or destructive process involving nerve axons)

   1. **Wallerian degeneration** — change in nerve fibers occurring **distal to site of a focal destructive lesion** of axons or following destruction of neuronal cell body (soma) itself (such as with trauma or ischemia)

      a. Early axonal and myelin fragmentation

      b. Subsequent Schwann cell proliferation and myelin phagocytosis

      c. Eventual migration of lipid-filled macrophages, and development of interstitial (endoneurial) fibrosis

      d. Formation of Schwann cell "tubes" (composed of stacks or rows of proliferated Schwann cells within original surrounding "tube" of basal lamina), through which regrowth of axon from proximal stump is possible

   2. **Distal axonal degeneration** — indolent process in variety of nerve diseases (such as inherited, toxic, or metabolic) characterized by gradual distal-to-proximal (so-called **"dying back"**) axonal break-up, associated with secondary myelin destruction

   B. Primary myelin degeneration (**demyelination**) — segmental process involving destruction of individual myelin internodes and/or Schwann cells; **axonal integrity preserved**; surviving Schwann cells proliferate and **remyelinate** axon

   C. **Axonal regeneration** — proximal intact axons in nerve stump attempt regeneration by sending out axonal sprouts to make contact with Schwann cell "tubes"

   1. **Neuroma** — results from aberrant axonal regeneration with sprouts unable to find Schwann cell "tubes" in which to regrow; tangled mass of axonal sprouts; often painful

D. Myelin regeneration (**remyelination**) — follows either primary demyelination or after axonal regeneration; occurs over previously demyelinated segments of intact axons or over naked regenerating axons; surviving Schwann cells elaborate new myelin sheath

E. **"Onion bulb" formation** — characteristic morphologic change in certain **chronic demyelinating or hereditary neuropathies**; repeated cycles of demyelination and remyelination result in **concentric layers of Schwann cells** and their processes around central axon (resembling onion bulbs in cross section)

F. **Nerve trauma**

1. **Neurapraxia** — nerve is injured (ie, compression) **without physical axonal disruption**, leading only to transitory "physiologic" dysfunction

2. **Axonotmesis** — **axon is disrupted** (ie, crushed) but Schwann cell tubes remain to facilitate axonal regeneration and substantial recovery

3. **Neurotmesis** — **completely severed nerve** such that amount of recovery depends on adequacy of alignment of fascicles in reanastomosis

VII. Common polyneuropathies

A. **Guillain-Barré Syndrome** (Landry-Guillain-Barré-Strohl syndrome; inflammatory polyradiculoneuropathy; ascending polyradiculoneuropathy)

1. Progressive **symmetric motor weakness** (usually distal > proximal and legs > arms) with paresthesias and loss of reflexes; cranial nerves ultimately involved in over 75% of cases; respiratory muscle weakness common

2. Many patients have antecedent viral illness, surgery, or immunization; possibly as many as 25% have preceding *Campylobacter jejuni* diarrheal illness

3. Potentially fatal autonomic dysfunction (including cardiac arrythmia, bladder disturbance, fluctuating blood pressure with hypotension, anal sphincter weakness, gastrointestinal motility disturbances, and sluggish pupils)

---

**SELECTED CAUSES OF POLYNEUROPATHY**

Acrylamide poisoning
Amyloidosis
Arsenic poisoning
Chloroquine therapy
Cryoglobulinemia
Diabetes mellitus
Diphtheria
Gold therapy
Isoniazid therapy
Lead poisoning
Liver failure
Mercury poisoning
Metachromatic leukodystrophy
Multiple myeloma
Myxedema
Nitrofurantoin therapy
Porphyria
Pyridoxine deficiency
Renal failure
Rheumatoid arthritis
Scleroderma
SLE
Tangier disease
Triorthocresyl phosphate poisoning
Vincristine therapy
Vitamin $B_{12}$ deficiency
Waldenström's disease

4. Elevated cerebrospinal fluid (CSF) protein with few cells (**albuminocytologic dissociation**)

5. Prolonged F-wave latency, slowing of nerve conduction velocities, and absent H-reflex.

6. **Autoimmune**-mediated demyelination with progression over 1-3 weeks followed by resolution (remyelination) over months to several years

7. Treatment:

    a. Monitoring of breathing and providing ventilatory support if necessary during initial acute progressive phase; in some cases, plasmapheresis during first week may shorten disease course

    b. During recovery, rehabilitation support is necessary; corticosteroid therapy is useful in those patients with relapsing form (chronic relapsing polyradiculoneuropathy)

B. **Diabetic polyneuropathy**

1. Most **common cause of polyneuropathy** in North America; some degree of peripheral polyneuropathy (sensory > motor) is evident in almost all diabetics

2. **Loss of sensation in feet** and to lesser extent hands ("stocking-glove distribution") along with paresthesias or hyperesthesias; superficial trauma may go unnoticed due to lack of sensation; when severe, loss of pain and proprioception (joint position sense) may lead to joint destruction (diabetic pseudotabes)

3. **Autonomic neuropathy** — changes in **sweating**, postural hypotension, impotence, **bladder dysfunction** (usually atonic bladder), and **disturbed gastrointestinal motility**

4. **Diabetic amyotrophy** — weakness and atrophy of **proximal leg muscles**

5. Must be differentiated from diabetic mononeuritis (or mononeuritis multiplex)

    a. Mononeuritis results from compromise of vasonervorum (nerve vasculature)

    b. Oculomotor nerve palsy — most common mononeuritis (followed by femoral nerve palsy); usually painful; diabetic oculomotor palsy presents with **eye deviated down and out but with pupillary reaction spared** (pupillary involvement suggests aneurysm of posterior communicating artery)

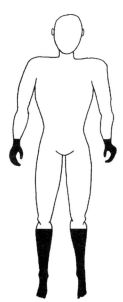

Stocking-glove sensory loss in diabetic polyneuropathy.

C. **Vitamin B$_{12}$ deficiency** (pernicious anemia; subacute combined degeneration; combined systems disease)

1. Symptoms of moderate to severe **loss of posterior column sensation** (proprioception; vibratory and position sense); on examination, patient falls from standing position after eye closure (**positive Romberg test**)

2. **Spasticity** (due to corticospinal tract involvement) with bilateral **extensor plantar responses** (Babinski reflexes) despite reduction or loss of tendon reflexes; can result in paraparesis or quadriparesis

3. Diagnosis established by finding low serum vitamin B$_{12}$ levels, elevated levels of methylmalonic acid in serum or urine, or **positive Schilling test**; anemia or disturbances of blood cell morphology not required for diagnosis

4. Treatment after diagnosis consists of parenteral vitamin B$_{12}$; administration of only folic acid to a patient with vitamin B$_{12}$ deficiency can worsen neurologic symptoms, while hematologic disturbances revert to normal

D. **Alcoholic polyneuropathy**

1. Presentation of **reduced sensibility in feet**, **absent ankle reflexes**, and paresthesias; motor symptoms minimal until late stage

2. Related to **thiamine deficiency** (probably in combination with other nutritional deficiencies) in chronic alcoholics

3. Treatment consists of cessation of alcohol abuse and restoration of adequate nutrition, along with **thiamine supplementation**

E. Other polyneuropathies

1. **Nutritional sensorimotor neuropathies**

    a. **Pyridoxine (vitamin B$_6$) deficiency**

       (1) Associated with **isoniazid** therapy for tuberculosis

       (2) Treatment consists of **vitamin B$_6$ supplementation**

    b. **Folic acid deficiency**

       (1) Associated with **phenytoin** treatment of epilepsy

(2) Treatment consists of **folic acid supplementation**

2. Toxic neuropathies

   a. **Arsenic poisoning** — **painful sensorimotor neuropathy** (red burning hands and feet); associated with anemia, brown skin discoloration, plantar and palmar hyperkeratosis, and white transverse bands in nails (**Mees' lines**); treated by chelation therapy with dimercaprol (BAL)

   b. **Lead poisoning** — mainly **motor neuropathy** mimicking mononeuritis multiplex (commonly **wrist drop** or **foot drop**); requires chronic exposure in adults (lead poisoning causes encephalopathy in children); associated with **anemia, basophilic stippling of red blood cells**, "lead line" along gingiva, and colicky abdominal pain; treated by chelation therapy with EDTA or penicillamine

   c. **Mercury poisoning** — dementia with **motor neuropathy**; acrodynia (Pink disease) in children; treated by chelation therapy with dimercaprol or EDTA

3. **Leprosy** — common neuropathy in underdeveloped countries due to **invasion of nerves** by acid-fast bacillus *Mycobacterium leprae*; nerves palpably **enlarged**

   a. **Tuberculoid leprosy** — **discrete** skin lesions with destruction of local nerves and sensorimotor deficit in distribution of destroyed nerves

   b. **Lepromatous leprosy** — after hematogenous spread of bacilli, **diffuse** infiltration of skin and peripheral nerves produces muscle weakness and widespread sensory loss (particularly in **cooler parts of body** such as pinnae of ears, tip of nose, and dorsum of feet and hands) and muscle weakness

VIII. **Motor neuropathies**

A. **Amyotrophic lateral sclerosis (ALS)**

   1. **Progressive generalized muscle weakness and wasting (atrophy) with fasciculations**; due to **death of anterior horn cells**

   2. **Upper motor neuron loss** results in spasticity with hyperactive tendon reflexes and extensor plantar responses (Babinski reflexes)

   3. No significant sensory abnormality

   4. Fatal from ventilatory failure generally within five years of onset

Tongue atrophy characteristic of amyotrophic lateral sclerosis.

5. Must be differentiated from treatable conditions:

    a. **Cervical spondylitic myelopathy**, **syringomyelia**, or spinal tumors identified by CT scan, MRI or myelography

    b. Parathyroid disease shows elevated serum calcium level

6. 20% of familial cases associated with gene locus on long arm of chromosome 21 resulting in defective copper/zinc-superoxide dismutase (SOD1)

7. Variant presentations

    a. **Progressive bulbar palsy** — primary involvement of brain stem and cranial motor nerves

    b. **Primary lateral sclerosis** — primary involvement of upper motor neurons with little anterior horn cell death

    c. **Progressive muscular atrophy** — primary involvement of anterior horn cells with little upper motor neuron loss

B. Inherited spinal muscular atrophy

1. **Werdnig-Hoffmann disease (infantile spinal muscular atrophy)**

    a. **Floppy infant** with progressive weakness, tongue fasciculations, feeding difficulties, and death from ventilatory insufficiency before age 2 years

    b. **Autosomal recessive disorder**

2. **Kugelberg-Welander disease** (juvenile spinal muscular atrophy)

    a. **Progressive proximal weakness** and fasciculations beginning in childhood or adolescence

    c. **Autosomal recessive** or **autosomal dominant** inheritance

IX. **Charcot-Marie-Tooth disease** (CMT1A, **peroneal muscular atrophy syndrome**, hereditary motor and sensory neuropathy)

A. Slowly progressive **distal muscle atrophy** due to progressive axonal loss; begins with **pes cavus** and hammer toe foot deformity and atrophy of lower legs ("stork legs" or "inverted champagne bottle legs"); sensory loss permits unnoticed injuries and ulceration; hands involved later in course

B. **Autosomal dominant** inheritance

1. Due to genetic abnormality on short arm of chromosome 17 (locus 17p11.2) involving peripheral myelin protein 22 (PMP22)

2. Duplication of DNA segment at this site is associated with CMT1A phenotype; deletion of this site is associated with hereditary neuropathy with liability to pressure palsies (tomaculous neuropathy)

C. In some families other neurologic signs may be present including palpably enlarged nerves, nystagmus, or cerebellar ataxia.

Pes cavus and hammer toe deformity typical of Charcot-Marie-Tooth disease.

X. **Poliomyelitis**

A. Acute **viral meningoencephalitis** with destruction of anterior horn cells, often preceded by **gastroenteritis**

B. Asymmetric, with flaccid weakness

XI. **Focal neuropathies, plexopathies, and radiculopathies**

A. **Carpal tunnel syndrome — compression** of **median nerve** by transverse carpal (volar) ligament at **wrist**

1. Pain and paresthesias in hand (especially over thumb and index finger), often worse at night and relieved by rubbing or shaking hand

2. Weakness and atrophy of **thenar muscles**; ultimately results in **"simian hand"** with inability to appose thumb to little finger

With light tap over median nerve, there is tingling and pain in thumb and index finger (Tinel's sign).

3. **Tinel's sign** — lightly tapping median nerve at wrist produces pain or tingling in thumb and index finger

4. Results from thickening of carpal ligament due to amyloid infiltration (multiple myeloma), rheumatoid arthritis, acromegaly, myxedema, or repeated trauma (may be occupation-related)

5. Nerve conduction studies show **prolonged distal latency** due to slowing of conduction in **compressed segment of median nerve**

## Chapter 9

      6. Treatment involves surgical release to relieve pressure on nerve

B. **Tardy ulnar palsy** — **compression** of **ulnar nerve** at condylar (ulnar) groove at **elbow**

      1. Progressive paresthesias of ring and little fingers

      2. Weakness of **finger extension** and abduction; muscle atrophy of **hypothenar eminence** and interossei (hollowing between metacarpal bones, especially evident in space between thumb and index finger); ultimately results in **clawhand** or "beer stein holder's hand" deformity

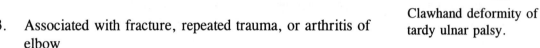

Clawhand deformity of tardy ulnar palsy.

      3. Associated with fracture, repeated trauma, or arthritis of elbow

      4. Nerve conduction studies show **slowing** of **ulnar nerve** conduction velocity **around elbow** (distal conduction velocity is normal)

      5. Treatment involves surgical relief of nerve compression or anterior transposition of nerve

C. **Radial nerve palsy**

      1. Weakness of extension of wrist and fingers (**wrist drop**)

      2. Usually due to pressure on radial nerve in axilla or upper arm: "Saturday night palsy" (drunken sleep with upper arm resting on edge of park bench) or "crutch palsy" (axillary pressure from wooden crutch)

      3. Treatment consists of preventing further pressure injury

D. **Klumpke-Déjérine palsy** — lesion of **lower trunk** (C8-T1) of brachial plexus

      1. Weakness of intrinsic hand muscles (clawhand deformity) and forearm flexors

      2. **Horner's syndrome** (unilateral **ptosis, miosis,** facial **anhidrosis**; lack of iris pigmentation, if congenital) occurs if damage includes T1 motor root containing sympathetic nerve fibers

      3. Reduced or absent triceps tendon reflex

Wrist drop from radial nerve palsy.

4. Results from sudden upward pull on shoulder (as during breech delivery or during fall) or from **apical lung tumor** infiltration **(Pancoast tumor)**

5. Spontaneous recovery usually occurs with traumatic lesion

E. **Erb-Duchenne palsy** — lesion of **upper trunk** (C5-C6) of brachial plexus

1. Weakness of shoulder and elbow resulting in arm dangling at side with fingers slightly flexed and palm facing backward (**"porter's tip position"**)

2. Absent biceps and brachioradialis tendon reflexes

3. Diffuse discomfort or pain in shoulder, but minimal sensory loss

4. Results from **forceful separation of head and shoulder**, such as occurs during difficult deliveries, from **motorcycle accidents**, during general anesthesia, or from carrying heavy backpack ("rucksack paralysis")

5. Must be differentiated from C6 root lesion which does not have shoulder weakness

6. Prognosis for spontaneous recovery good

F. **Brachial plexus neuritis** (brachial plexitis)

1. **Abrupt onset** of severe aching **shoulder and neck pain (often beginning at night)** made **worse by arm movement**; followed several days later by nearly complete **paralysis** involving mostly muscles innervated by upper trunk of brachial plexus

2. Reflexes variably reduced and sensory loss is minimal or not detectable

3. Bilateral involvement may occur, but dominant arm usually more affected

4. Spontaneous recovery may take up to two years and supportive physical therapy necessary to prevent contractures and shoulder arthropathy

G. **Cervical radiculopathy** — **root compression** from protruded cervical intervertebral disc

"Porter's tip position" of hand in lesion of upper trunk of brachial plexus (Erb-Duchenne palsy).

1. May present as **nerve root signs** alone or may be accompanied by signs of **spinal cord compression** (spastic paraparesis with hyperactive tendon reflexes and extensor plantar responses)

2. **Neck pain** and paravertebral **muscle spasm** common; pain may radiate into arm

3. Paresthesias, sensory loss, and weakness in distribution of compressed nerve root

   a. **C6 root** — compression by C5-C6 disc herniation

      (1) Sensory loss and paresthesias primarily of **thumb and index finger**

      (2) Weakness of **elbow flexion (biceps muscle)** and forearm pronation

      (3) Decreased **biceps** and **brachioradialis** tendon reflexes

   b. **C7 root** — compression by C6-C7 disc herniation

      (1) Sensory loss and paresthesias of **index and middle fingers**

      (2) Weakness of **elbow extension (triceps muscle)** and wrist extension

      (3) Decreased **triceps** tendon reflex

4. Myelography, CT scan, or MRI identify root compression

5. Treatment with rest, heat, and cervical traction may alleviate pain; neurologic signs of weakness or cord compression necessitate surgical decompression

H. **Bell's palsy** (idiopathic facial palsy)

1. Acute or subacute complete unilateral paralysis of **facial muscles** (both lower face and forehead), often preceded or accompanied by pain behind ear; preceding nonspecific viral illness common

2. Impaired taste, hyperacusis (from stapedius muscle paralysis), and reduced tearing often present

3. Must be differentiated from:

   a. **Central facial palsy** (lesion of upper motor neuron controlling facial movement) which spares forehead

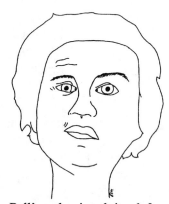

Bell's palsy involving left side of face with drooping of corner of mouth and widening of palpebral fissure.

b. **Cerebellopontine angle tumor** — usually accompanied by decreased hearing (vestibuloacoustic nerve involvement) or decreased corneal reflex (facial or trigeminal nerve involvement)

c. **Diabetic mononeuritis**

d. Chronic **inflammation** or infection of **ear** or **parotid gland** (including **sarcoidosis**)

4. Most patients spontaneously recover without treatment, but corticosteroids may reduce nerve swelling at onset and hasten recovery; **corneal protection** is necessary until lid closure becomes possible

I. **Trigeminal neuralgia (tic douloureux)**

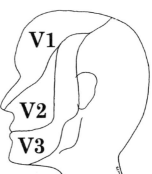

Sensory distributions of three divisions of trigeminal nerve (cranial nerve V).

1. Brief **paroxysmal lancinating pain** in distribution of **trigeminal nerve** (cranial nerve V) causing patient to grimace (tic); second and third divisions more commonly affected than first division

2. Paroxysms can be initiated by touching or moving **trigger points** on face

3. No identifiable weakness or sensory loss

4. Onset usually after age 40 years; women more commonly afflicted than men; etiology is unknown

5. Must be differentiated from **atypical trigeminal neuralgia** (due to **multiple sclerosis** or **posterior fossa mass lesion** such as tumor or aneurysm) which usually presents in younger individuals and has associated sensory loss in trigeminal distribution or other cranial nerve palsies

6. Treatment with carbamazepine relieves symptoms in most cases; refractory cases may require surgical ablation of gasserian (trigeminal) ganglion or vascular decompression of trigeminal nerve root entry zone

J. **Femoral nerve palsy**

1. Weakness and wasting of **quadriceps muscle** producing difficulty with leg extension; resultant problems include difficulty climbing stairs and compensation by walking with knee rigid; longstanding lesions result in development of genu recurvatum

2. Absent patellar (knee) reflex

3. Frequently associated with **diabetes mellitus** (diabetic mononeuritis) or inguinal trauma

4. Must be differentiated from disuse atrophy (knee reflex preserved) and polymyositis (elevated serum CPK)

K. **Peroneal nerve palsy**

1. **Foot drop** ("slapping foot") due to weakness of foot and toe extension and foot eversion (turning out); atrophy of anterior compartment of lower leg

2. Minimal sensory loss over anterolateral lower leg and dorsum of foot

3. Results from pressure on **common peroneal nerve** as it courses over **fibular head** laterally at knee; causes include prolonged crossing of legs while seated, pressure during sleep, coma, or anesthesia, or pressure by tight plaster casts, knee boots, or knee braces

4. Must be differentiated from L5 root lesion in which internal hamstring reflex can be absent and posterior tibial muscle (inverts plantar-flexed foot) is weak

L. **Meralgia paresthetica** — compression of **lateral femoral cutaneous nerve** (pure sensory nerve) by inguinal ligament

1. **Burning paresthesias** and loss of sensation on lateral aspect of thigh

2. Associated with obesity, pregnancy, diabetes, inguinal trauma, or use of tight-fitting corset

M. **Lumbosacral radiculopathy** — root compression from protruded intervertebral disc

1. Nerve root signs produced by protrusion of intervertebral disc at one higher level

2. Symptoms often begin following trauma

3. Low back pain and paravertebral muscle spasm result in **pelvic tilt**; **pain radiates into leg**, is aggravated by activity and is relieved by bed rest; **positive straight leg raising test** (with patient lying prone, pain elicited by passive raising of straight leg)

4. Paresthesias, sensory loss, and weakness in distribution of compressed nerve root

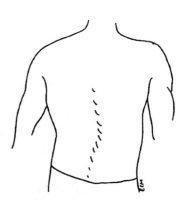

Pelvic tilt from paravertebral muscle spasm in lumbar radiculopathy.

# PERIPHERAL NERVOUS SYSTEM DISORDERS

    a. **S1 root** — compression by L5-S1 disc herniation

       (1) Sensory loss and paresthesias over lateral ankle, heel, and sole

       (2) Calf muscle (gastrocnemius) weakness with difficulty standing up on toes in affected leg

       (3) Decreased or absent Achilles tendon reflex (ankle reflex)

    b. **L5 root** — compression by L4-L5 disc herniation

       (1) Paresthesias and sensory loss and paresthesias over anterolateral lower leg and dorsum of foot and big toe

       (2) **Foot drop** with weakness of dorsiflexion and eversion of foot

       (3) Diminished internal hamstring tendon reflex

    c. **L4 root** — compression by L3-L4 disc herniation

       (1) Sensory loss and paresthesias over shin and medial lower leg to medial ankle and foot

       (2) Weakness of **quadriceps muscle**

       (3) Decreased or absent patellar tendon (knee) reflex

Passively lifting and straightening leg and knee produces pain radiating into leg.

5. Myelography, CT scan, or MRI identify root compression

6. Treatment with bed rest, heat, and lumbar corset may alleviate pain; neurologic signs of weakness necessitate surgical decompression

XII. **Neuromuscular junction diseases**

  A. **Myasthenia gravis**

    1. Progressive **weakness after exercise ("fatiguable weakness")** with most severe involvement of **ocular muscles**, followed by **bulbar** and respiratory muscles, while extremity muscles are least affected

a. Simulates effect of **partial curarization** (exposure to low doses of nondepolarizing muscle relaxant *d*-tubocurarine)

b. Weakness can be exacerbated by drugs interfering with neuromuscular transmission, including succinylcholine, quinidine, procainamide, quinine (tonic water), penicillamine, streptomycin, kanamycin, polymyxin, lincomycin, tetracycline, gentamicin; safe drugs include penicillin, cephalothin, rifampin, vancomycin, amphotericin, nystatin

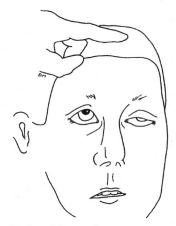

Fatiguable weakness in myasthenia gravis: during prolonged upgaze, eyelid droops (evident here on patient's left side) and eye deviates downward.

2. Initial presentation with **ocular findings** (diplopia or ptosis) usually progresses within one year to bulbar and extremity muscle weakness; some patients never progress beyond ocular involvement (**ocular myasthenia**)

3. **Autoimmune** attack on muscle endplate acetylcholine receptors, with serum **anti-acetylcholine receptor (antiAChR) antibodies (IgG)** detectable in most patients; other simultaneous autoimmune diseases common

4. Jolly test — progressive **decrement** in amplitude of muscle action potential recorded during 3 Hz repetitive nerve stimulation

5. **Tensilon test** — intravenous injection of short-acting **acetylcholinesterase inhibitor** edrophonium (Tensilon) results in almost immediate dramatic improvement in strength lasting for several minutes

6. Females more commonly affected than males; peak incidences in younger women (ages 15-30 years) and older men (ages 50-70 years)

7. Frequently associated with **thymic abnormalities**, particularly **hyperplasia** in young women and **thymoma** in old men

8. **Transient neonatal myasthenia gravis** — some babies born to myasthenic mothers have transient feeding difficulties, weak cry, breathing difficulties, and floppiness; caused by **transplacental passage of antiAChR antibodies**; supportive treatment necessary until spontaneous recovery during first week of life

9. Treatment includes anticholinesterases to augment acetylcholine neurotransmission (pyridostigmine), immunosuppressive drugs (prednisone or **azathioprine**), plasmapheresis

(to remove antibodies), and thymectomy; sometimes breathing difficulties necessitate emergency ventilatory support

B. **Lambert-Eaton (myasthenic) syndrome**

1. Progressive **generalized muscle weakness** (proximal greater than distal) that **improves with exercise**; absent tendon reflexes and mild sensory loss also evident

2. Ocular and bulbar musculature are spared, but autonomic symptoms (ptosis, impotence, dry mouth, constipation) present

3. **Autoimmune** attack on presynaptic nicotinic cholinergic nerve terminals, with IgG antibodies detectable in most patients

4. **Incrementing response** of muscle action potential recorded during high frequency repetitive nerve stimulation

5. Commonly associated with **carcinoma**, particularly **small cell (oat cell) lung carcinoma**, or with other autoimmune diseases

6. Treatment is directed at underlying malignancy or autoimmune disease; immunosuppressive drugs, plasmapheresis, or drugs that augment neurotransmitter release and synaptic function (guanidine or 3,4-diaminopyridine) can increase strength

C. **Botulism**

1. Food-borne (classical) botulism

    a. Acute, rapidly progressive **paralysis of extraocular and bulbar** (pharyngeal) muscles; skeletal muscle weakness and respiratory compromise follow within 2 to 3 days; **constipation** occurs from smooth muscle paralysis

    b. Can be distinguished from myasthenia gravis by presence of **nonreactive dilated pupils** and from Guillain-Barré syndrome by lack of sensory findings and normal cerebrospinal fluid

    c. Caused by ingestion of preformed **exotoxin** (which blocks acetylcholine release from nerve terminals) produced by anaerobic gram-positive spore-forming rod bacteria *Clostridium botulinum*, most often from eating improperly canned vegetables

    d. Diagnosis confirmed by repetitive nerve stimulation studies or by assay of toxin in serum or tainted food

e. Treatment includes ventilatory support for respiratory failure, administration of horse serum antitoxin, and use of guanidine to augment neurotransmitter release

2. **Neonatal botulism**

   a. Self-limited disorder occurring in infants whose gastrointestinal tracts are colonized by *Clostridium botulinum* which produce exotoxin that is subsequently absorbed

   b. Characterized by floppiness and ventilatory failure

   c. Treatment involves antibiotics and ventilatory support

XIII. **Muscle diseases**

A. **Polymyositis**

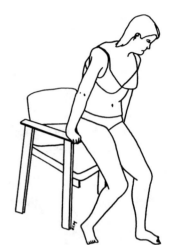

Proximal muscle weakness in polymyositis necessitates pushing off with arms to rise from sitting position.

1. Subacute onset of progressive **proximal muscle weakness**, resulting in difficulty walking stairs or getting up from chair; occasionally associated with muscle discomfort or pain

2. Sometimes associated with **skin involvement (dermatomyositis)** including purple discoloration of eyelids, skin erythema (dorsal surfaces of hands and over finger, elbow, and knee joints), and subcutaneous calcifications.

3. Serum muscle enzymes (creatine phosphokinase and aldolase) elevated; muscle biopsy shows **inflammation** and muscle fiber necrosis

4. **Autoimmune** attack on skeletal muscle fibers; linked to occult cancers and to other autoimmune disorders

5. Treatment involves use of immunosuppressive drugs (including prednisone, azathioprine, or methotrexate)

B. **Duchenne muscular dystrophy**

1. Hereditary disorder **(X-linked transmission)** presenting initially around age 3 years as waddling gait and difficulty climbing stairs; clinical examination reveals **Gowers' maneuver** and enlarged calf muscles **(pseudohypertrophy)** which have rubbery consistency

2. Progressive **proximal weakness** results in inability to stand by age 12 years necessitating wheelchair confinement; subsequent contractures, kyphoscoliosis, and further weakness usually result in death from pulmonary infections around age 20 years

3. Intellectual impairment is common

4. Serum creatine phosphokinase (CPK) is greatly elevated (often 100 times normal) and muscle biopsy reveals fiber destruction with marked fibrosis

5. Results from complete absence of muscle membrane protein **dystrophin** normally produced by X chromosome gene (Xp21 gene locus); **partial deficiency** of dystrophin results in milder phenotype (**Becker muscular dystrophy**); occasional females carrying abnormal gene manifest muscle symptoms due to **lyonization** (inactivation) of normal X chromosome

Gowers' maneuver in Duchenne muscular dystrophy: boy must climb up on himself to come to standing position.

6. Treatment involves physical therapy to prevent contractures, bracing to maintain ambulation, respiratory care to prevent pulmonary infections, and genetic counseling (including identification of female carriers by DNA analysis)

C. **Myotonic muscular dystrophy**

1. **Myotonia** - muscular stiffness from repetitive muscle fiber contraction

   a. After tightly clenching fist for 30 seconds, patient has difficulty in opening hand

   b. Percussion of thenar eminence with reflex hammer causes thumb to move up to oppose little finger and remain in this position for several seconds

   c. Percussion of gastrocnemius muscle produces transient hard lump

   d. Electromyography (EMG) reveals characteristic **"dive-bomber" potentials**

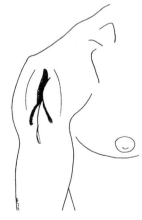

Myotonia is evident as prolonged dimpling of muscle after direct blow from reflex hammer.

2. Slowly progressive distal weakness and wasting, especially in upper extremities

3. **Multisystem involvement** with **cardiac arrythmia**, **cataracts**, frontal baldness, abnormal gastrointestinal motility, endocrine dysfunction, and intellectual and behavioral disturbances; widely varying spectrum of severity and age of clinical onset of symptoms

4. Most common genetically-determined muscle disease

   a. **Autosomal dominant inheritance pattern** due to defective gene locus on chromosome 19

   b. Genetic abnormality consists of expansion (increased length) of unstable trinucleotide (CTG) repeat sequence in gene encoding for myotonin-protein kinase

   c. Anticipation — increasing severity of symptoms over generations; relates to increasing length of trinucleotide repeat sequence which correlates with earlier onset of symptoms

5. May present in infancy as **floppy baby** with respiratory distress and difficulty in feeding; occurs in infants with myotonic dystrophy born to **mothers with myotonic dystrophy**

6. Myotonia also occurs in myotonia congenita and hyperkalemic periodic paralysis, but without systemic symptoms

7. Treatment involves use of drugs which interfere with muscle membrane depolarization (quinine, procainamide, phenytoin), insertion of cardiac pacemaker to counteract arrhythmias, and genetic counseling

D. **Type 2 muscle fiber atrophy** — muscle weakness associated with systemic disease

1. Subacute or chronic development of **proximal muscle weakness**

2. Characteristic **selective atrophy of type 2 muscle fibers**, identifiable only by histopathologic study of frozen muscle biopsy specimens

3. Electromyography (EMG), nerve conduction studies, and serum levels of creatine phosphokinase (CPK) normal

4. Associated with disuse (**"disuse atrophy"**) or with various chronic disorders such as **cachexia**, hypercorticism (**"steroid myopathy"**), hyperparathyroidism, or hyperthyroidism (**"thyroid myopathy"**); may complicate treatment of various medical disorders

5. Treatment is directed at underlying illness along with exercise program to increase muscle strength

E. **Malignant hyperthermia**

1. Sudden onset of marked **hyperthermia** (body temperature of 42°C or higher) during general **anesthesia** with **halogenated anesthetics** (such as halothane or enflurane) or **succinylcholine**; additional symptoms include extreme muscular **rigidity**, **hyperkalemia**, tachycardia, tachypnea, severe metabolic and respiratory **acidosis**, and **myoglobinuria**

2. Results from genetic defect in sarcoplasmic reticulum; massive release of normally sequestered calcium ions produces intense muscle contraction

3. **Emergency treatment** is necessary, including immediate cessation of anesthetic agents, intravenous administration of **dantrolene**, and initiation of cooling procedures

4. Caffeine contracture test on muscle biopsy specimen can identify susceptible family members

5. Susceptible individuals can occasionally undergo anesthesia using known triggering agents without inducing attack

F. **Volkmann's ischemic contracture**

1. Swelling of injured forearm tissue may occlude blood vessels causing acute ischemia and **infarction** of muscles and nerves with subsequent **fibrosis** and contracture

Volkmann's ischemic contracture.

2. Prevention of ischemia by **urgent surgical fasciotomy** to relieve pressure is usually necessary

G. **Myositis**

1. **Localized muscle inflammation** secondary to **bacterial infections** (*Staphylococcus* or *Streptococcus* abscess), **parasitic infestation** (toxoplasmosis, **trichinosis**, *Toxocara*), or **granulomatous disease** (**tuberculosis**, histoplasmosis, actinomycosis, sarcoidosis)

2. Subacute or acute muscle **pain**, **swelling**, and weakness, with elevation of serum creatine phosphokinase (CPK)

3. Identification of causative agent requires muscle biopsy and culture

4. Treatment is directed at eliminating specific causative organism; with marked swelling, fasciotomy may be necessary to prevent vascular compromise and consequent ischemic necrosis

# Chapter 10                                                  DIZZINESS AND VERTIGO

I. Terminology

    A. **Dizziness** — **imprecise term** used by patients to describe various sensations ranging from whirling to lightheadedness to unsteadiness

    B. **Vertigo** — **illusion** (sensation) of rotational **movement of self or environment**; described as whirling, spinning, tilting, rotating; often accompanied by nystagmus, unsteadiness, and nausea; implies abnormality of peripheral vestibular apparatus, vestibular nerve, or central nervous system (brain stem vestibular pathways)

    C. **Syncope** — loss of consciousness or **fainting**; often accompanied by profuse sweating (diaphoresis) and nausea; due to **reduced cerebral perfusion** and usually implies cardiovascular abnormality

    D. **Dysequilibrium** — unsteadiness ("drunkenness") without vertigo; implies **incongruity of sensory inputs mediating spatial orientation**

        1. Feeling experienced by individual who has lost proprioceptive input from legs (due to peripheral neuropathy) and cannot see foot position when attempting to walk in darkened room

        2. Elderly individuals often have **altered sensory input** due to reduced vision (cataracts), reduced proprioception (as from diabetic peripheral polyneuropathy), or vestibular abnormalities (due to ischemia from basilar artery insufficiency)

        3. Diminished sensory input causes feeling of being out of touch with environment, especially when walking or turning

        4. Treatment involves supplying additional sensory cues (such as using cane when walking)

    E. **Nystagmus** — involuntary **rhythmic oscillation of eye**; evident as slow drift of eye in one direction followed by quick (cerebral cortex controlled) corrective jerk in opposite direction; **named for direction of quick (fast) component**

        1. **Toxic-metabolic** disorders produce **symmetrical horizontal nystagmus**, equal in both directions of gaze; common with large doses of **phenytoin** (Dilantin), **benzodiazepines** such as diazepam (Valium), or **barbiturates**

2. **Asymmetric nystagmus** - absent or **reduced in one direction of gaze** compared with opposite direction of gaze; usually indicates **peripheral vestibular dysfunction** or **brain disease**

3. **Dysconjugate nystagmus** — movements **greater in one eye**; always indicates **brain disease**

4. **Upward-gaze, downward-gaze, or rotatory nystagmus** — movements other than in horizontal plane suggest **brain disease** and are usually not associated with vertigo

5. **Positional nystagmus** — nystagmus (and associated vertigo) occurring with certain head positions or with head movement

F. **Oscillopsia** — **illusion** that stationary objects (such as ground, buildings, or walls) are **moving back and forth** (oscillating); results from **bilateral vestibular damage** (as from **vestibulotoxic drugs**) or **brain stem disease** (as in **lateral medullary infarction**)

II. Functional testing of **vestibuloacoustic nerve (cranial nerve VIII; auditory or acoustic nerve; vestibular nerve)**

A. **Weber's test** — with vibrating tuning fork placed in **midline on patient's forehead**, sound should be heard equally in both ears; with unilateral damage to cochlea or cranial nerve VIII (**sensorineural deafness**), sound is heard better in ear with normal acuity; with middle ear or outer ear disease (**conduction deafness**), sound is heard better in diseased ear

B. **Rinne test** — vibrating tuning fork first is held in front of external auditory meatus (**air conduction**) and then stem of tuning fork is placed firmly against mastoid process (**bone conduction**); bone conduction should be louder than air conduction with **conductive deafness** (middle or outer ear disease); air conduction is louder than bone conduction in normal individuals and those with **sensorineural deafness**

C. **Nylen-Bárány maneuver** — patient seated on examining table is suddenly lowered to supine position with head thrust 45 degrees backward over end of table and turned 45 degrees to one side; development of nystagmus (particularly asymmetric) or vertigo suggests vestibular disease

D. **Caloric test** — **unilateral vestibular stimulation** is accomplished by instillation of cold or warm water into one external auditory meatus with production of asymmetric nystagmus; test is performed with patient lying down and head flexed 30 degrees in order to stimulate **horizontal semicircular canal**

1. Warm water — slow movement of eyes away from stimulated ear with fast correction toward stimulated ear

Chapter 10

2. Cold water — slow movement of eyes toward stimulated ear and fast correction away from stimulated ear

3. Mnemonic for direction of nystagmus (named for fast corrective component) produced by caloric test — *COWS* (*c*old *o*pposite, *w*arm *s*ame)

4. Effect of **cold water** is similar to **destructive lesion** of vestibular apparatus or vestibular nerve; effect of **warm water** is same as **irritative lesion**

E. **Electronystagmometry** — electrodes placed around eye provide sensitive means of recording eye movements of nystagmus (either spontaneous or induced by testing)

F. **Audiometry** — hearing tests

1. **Pure tone audiometry** — determines **loudness threshold** for hearing sounds of various frequencies; **low frequency** hearing loss suggests **conductive deafness**, while **high frequency** hearing loss indicates **sensorineural deafness**

2. **Speech discrimination** — impaired ability to understand words despite normal (or nearly normal) pure tone audiometry; indicates **retrocochlear lesion (lesion of auditory nerve or brain stem)**

3. **Alternate binaural loudness balance (ABLB)** — test of **recruitment** in which ear with hearing loss cannot detect low intensity sounds but hears loud sounds equally with good ear; indicates **cochlear disease**

4. **Short increment sensitivity index (SISI)** — Determines ability to detect **small increases in loudness** while listening to simultaneous continuous tone; these sound increments are not heard with **retrocochlear lesion (lesion of auditory nerve or brain stem)**

5. **Tone decay** — continuous tones become **progressively inaudible** with **auditory nerve lesion**, while there is no change in loudness in normal individuals or with cochlear disease

6. **Impedance tympanometry** — measurement of ability of tympanic membrane to move in response to varying air pressures; excess movement suggests disruption of ossicular chain, while **reduced movement** suggests **otosclerosis** or **middle ear fluid collection** (effusion or infection) [particularly useful in evaluating conductive hearing loss in young children with repeated ear infections]

7. **Stapedial reflex** — protective reflex in which **loud noise** in one ear normally causes stapedius muscles in both ears to contract; **reflex arc** consists of **auditory nerve** (afferent limb of reflex), **brain stem interneurons**, and **facial nerve** (efferent limb of reflex); lesions of **auditory nerve** often reduce or abolish reflex even when hearing is still normal

8. **Békésy audiograms** — measure of perceived differences in loudness of continuous versus pulsed tone; type I tracing is normal, type II tracing suggests cochlear disease, while **type III** and **type IV** tracings indicate **retrocochlear lesion (lesion of auditory nerve or brain stem)**

9. **Brain stem auditory evoked response (BAER)** — computerized averaging of scalp potentials produced by click sounds in each ear

    a. Changes in **latency** or **amplitude** of recorded potentials suggests disease in peripheral or central auditory pathways

    b. **Potential peaks** presumably originate from specific sites in auditory pathway and latency between peaks (**interpeak latency**) is indicative of intactness of connecting tracts: wave I originates from auditory nerve, wave II from cochlear nuclei in upper medulla, wave III from trapezoid bodies and superior olivary nuclei bilaterally in lower pons, wave IV in lateral lemniscus and its nuclei in upper pons, wave V in midbrain inferior colliculi, wave VI in medial geniculate nuclei, and wave VII in primary auditory cortex (Heschl's gyri).

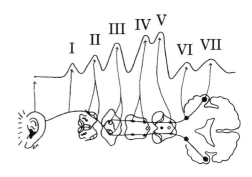

Representation of BAER potential peaks and responsible auditory pathway.

III. **Vertiginous disorders**

  A. **Benign positional vertigo**

   1. **Sudden paroxysmal vertigo** precipitated by assuming certain **head positions** or by **head motion** (positional vertigo)

   2. Episodes are **brief** (often associated with nausea) and recur frequently over days or weeks followed usually by **spontaneous remission**; occasional patients have relapses after months or years

   3. Distinguished from other causes of positional vertigo by repetition of Nylen-Bárány maneuver: with repetitions, vertiginous symptoms and nystagmus become less apparent (**fatigue of response**) in benign positional vertigo, but remain constant with other causes of positional vertigo

   4. **Hearing is normal**

---

**VERTIGINOUS DISORDERS**
**Vestibular Abnormality**
  Benign positional vertigo
  Ménière's disease
  Motion sickness
  Post-traumatic vertigo
  Toxic labyrinthitis
  Vestibular neuronitis
**Brain Stem Abnormality**
  Basilar artery migraine
  Brain stem strokes
  Multiple sclerosis
  Posterior fossa tumors

5. Treatment of symptoms necessary with antivertigo (anti-motion sickness) drugs until spontaneous remission occurs

B. Ménière's disease

1. **Recurrent sudden attacks of severe vertigo** lasting minutes to hours; during episodes patient is usually unable to stand or walk

2. Attacks are associated with **tinnitus (ringing in ear)**, nausea and vomiting, and feeling of ear fullness or pressure; **decrement in hearing** associated with each attack

| ANTIVERTIGO DRUGS |
|---|
| **Anticholinergics** |
|   Scopolamine (Transderm Scōp) |
| **Antihistamines** |
|   Dimenhydrinate (Dramamine) |
|   Diphenhydramine (Benadryl) |
|   Meclizine (Antivert) |
| **Phenothiazines** |
|   Promethazine (Phenergan) |
| **Sympathomimetics** |
|   Ephedrine |
| **Tranquilizers** |
|   Hydroxyzine (Vistaril) |

3. Disease has **fluctuating course** with attacks occurring up to many times per week for several months followed by long periods of remission; hearing loss may also fluctuate between attacks, but ultimately there is **complete sensorineural deafness** (at which time **vertiginous attacks cease**)

4. Onset usually after age 40 years and both sexes affected equally

5. Etiology is unknown; pathologic changes consist of **dilation of endolymphatic system (endolymphatic hydrops)** and **degeneration of hair cells**

6. Treatment is mainly symptomatic with bedrest and antivertigo medication; attempts at prevention of attacks with diuretics, low-sodium diets, and vasodilators have no proven value; surgical ablation of vestibular organ has been used in some patients with severe frequent attacks

C. Vestibular neuronitis (vestibular neuropathy; acute labyrinthitis)

1. **Acute onset of vertigo and nystagmus** accompanied by nausea and vomiting, that peaks in about one day and spontaneously subsides within one to three weeks

2. Patient is extremely apprehensive and attempts to **remain motionless**

3. Sense of fullness may be present in abnormal ear, but there is **conspicuous absence of tinnitus or alteration in hearing**

4. Nystagmus has **slow phase toward abnormal ear**; caloric response minimal or absent in abnormal ear

5. Often associated with **viral nasopharyngitis** and presumably results from virus-related damage to vestibular apparatus

6. Treatment involves reducing vertiginous symptoms with antivertigo drugs until spontaneous recovery occurs

D. **Toxic labyrinthitis** — numerous common **drugs** can damage vestibular organ, cochlea, or both

E. **Post-traumatic vertigo** — can result from **basilar skull fractures** that pass through **petrous pyramid** damaging delicate vestibular apparatus or from acceleration-deceleration injury (without fracture)

F. **Motion sickness** — some individuals have increased vestibular system **sensitivity to motion** resulting in vertigo (along with nausea and vomiting); adults with **migraine** frequently have history of motion sickness during childhood; **prophylactic treatment** with antivertigo drugs prevents symptoms when travel is necessary

| VESTIBULOTOXINS |
|---|
| **Alkaloids** |
|   Quinidine |
|   Quinine |
| **Aminoglycosides** |
|   Gentamicin |
|   Streptomycin |
|   Tobramycin |
| **Anticonvulsants** |
|   Phenytoin |
| **Anti-inflammatory** |
|   Phenylbutazone |
| **Polymyxin-B** |
| **Diuretics** |
|   Furosemide |
| **Salicylates** |

G. **Acoustic neuroma (acoustic neurilemoma or schwannoma)**

1. Insidious unilateral **hearing loss, tinnitus**, and **unsteadiness** with vertigo developing late in course of disease

2. Posterior fossa tumor that grows out of **internal auditory meatus** into **cerebellopontine angle** compressing adjacent cranial nerves (symptoms of facial numbness and weakness) and cerebellum (resulting in clumsiness or ataxia)

3. Associated with **café-au-lait spots** in **neurofibromatosis**

4. Diagnosis made with audiometry, brain stem auditory evoked responses, and radiologic imaging studies; surgical resection is necessary

H. **Multiple sclerosis** — demyelination in brain stem can damage central vestibular pathways leading to nystagmus and vertigo; more often, patients have **dysequilibrium** from disturbances in sensory input due to demyelinating plaques involving optic nerve and spinal cord

I. **Basilar artery migraine** — variant of classical migraine usually presenting in **children and adolescents**; aura relates to ischemia in distribution of basilar artery resulting in vertigo along with visual disturbance, ataxia, and impaired consciousness

J. **Brain stem ischemia** — **vertebrobasilar artery insufficiency** frequently presents as vertigo along with other signs and symptoms attributable to **brain stem dysfunction**; transient symptoms indicate reversible vascular compromise, while permanent signs or symptoms indicate infarction

1. **Subclavian steal syndrome**

    a. Attacks of **vertigo, ataxia**, or **unsteadiness** and **lightheadedness**, especially upon **exercising arm** on affected side; arm pain, fatigue, and cramping can occur during exercise

    b. Results from **occlusion of subclavian artery proximal to origin of vertebral artery** with consequent **reversal of blood flow (shunting** or "**steal**") from cranium down extracranial vertebral artery to supply ischemic arm

    c. Surgical correction is necessary to avoid brain stem infarction

2. **Lateral medullary infarction (Wallenberg syndrome)**

    a. Acute onset of severe **vertigo, nausea, vomiting, nystagmus**, and **oscillopsia**

    b. Examination reveals **gait and ipsilateral limb ataxia**, loss of **pain and temperature sense on ipsilateral face** (decreased corneal reflex) and **contralateral body**, **ipsilateral weakness of palate and vocal cords** with diminished gag reflex, **dysphagia**, hoarseness, and **ipsilateral Horner's syndrome** (ptosis, miosis, and facial anhidrosis); **intractable hiccups** may also occur

    c. *No* **limb weakness or paralysis**

    d. Infarction of **dorsolateral medulla** in distribution of **posterior inferior cerebellar artery (PICA)**; due either to **occlusion of vertebral artery** proximal to origin of PICA or (less commonly) to occlusion of PICA

    e. Spontaneous resolution of vertigo commonly occurs and other symptoms tend to improve

IV. **Hyperventilation syndrome** — common cause of **lightheadedness** (dizziness); associated with **circumoral paresthesias, paresthesias of fingers**, and **carpopedal spasm (cramps in hand and foot muscles)**; related to anxiety and depression

V. **Syncope and fainting**

   A. In upright position, individual has **prodrome** of uneasiness (queasiness), apprehension, **headache, facial pallor** ("ashen-gray color"), **profuse sweating, nausea** (sometimes vomiting), and **visual blurring** ("gray-out"), followed by loss of consciousness (slumping to ground), unless individual is able to lie down (which aborts loss of consciousness)

   B. Results from **reduction of blood flow to brain** (usually blood pressure less than 60 mm Hg); once individual is horizontal, normal blood flow to brain is restored and arousal occurs

C. Duration of unconsciousness varies; if individual is maintained in upright position, brain perfusion is not adequately restored, unconsciousness continues and after about 15 seconds convulsive activity occurs (**convulsive syncope**) lasting several minutes; differentiated from epileptic seizure by lack of postictal confusion and sleepiness

D. **Vasovagal syncope**

   1. "Common **fainting**" resulting from **sudden peripheral vasodilation** (particularly involving intramuscular arterioles); frequently caused by emotional stress, warm crowded conditions, or pain

   2. **Breath-holding spells**

      a. Form of vasovagal syncope in children aged 6 months to 6 years; child will lose consciousness (usually lasting less than 60 seconds), often followed by few seconds of tonic stiffening at end of period of unconsciousness (may be mistaken for seizure)

      b. **Cyanotic** type — begins with crying; **child holds breath, turns blue,** and **loses consciousness**, after which breathing resumes

      c. **Pallid** type — following minor startle or sudden painful stimulus, child briefly cries out, turns white (pale), and loses consciousness

      d. Benign condition that is very frightening to parents who should be reassured of benign prognosis: child will outgrow disorder and it does not lead to epilepsy

E. Other causes include **orthostatic (postural) hypotension** (such as hypovolemia, diabetic autonomic neuropathy, or antihypertensive medication treatment), **cardiac arrhythmias** (such as **Stokes-Adams attacks** due to prolonged asystole from atrioventricular block), and **carotid sinus hypersensitivity** (hypotension or bradycardia induced by minor stimulation of carotid sinus, such as by head turning or tight collar)

# Chapter 11　　　　　　　　　　CENTRAL NERVOUS SYSTEM NEOPLASMS

I. Pathophysiology of symptoms

   A. Local effects of tumors

   1. **Infiltration**, **invasion**, and **destruction** of normal central nervous system tissues by tumor produces focal neurological signs

   2. Mass of tumor produces direct pressure on neural structures causing degeneration (although brain and spinal cord can adjust remarkably to gradually increasing pressure)

   3. **Compromise of local circulation** due to direct pressure on capillaries and small arteries and veins can be associated with local tissue necrosis or more distant infarction

   4. Brain **edema** — usually greatest around tumor; can interfere with functioning of more remote neural tissue, adding to clinical symptoms directly attributable to tumor mass and eventually (usually over period of months) resulting in demyelination and astrocytic hyperplasia in white matter

      a. Treatment with high doses of **corticosteroids** can reduce edema

| PRINCIPAL BRAIN TUMORS BY LOCATION | |
|---|---|
| **Cerebral hemisphere** | **Fourth ventricle** |
| Astrocytoma | Choroid plexus papilloma |
| Ependymoma | Ependymoma |
| Meningioma | Meningioma |
| Metastatic carcinoma | **Pituitary region** |
| Oligodendroglioma | Craniopharyngioma |
| Vascular malformation | Pituitary adenoma |
| **Optic chiasm** | Meningioma |
| Astrocytoma | **Brain stem** |
| Meningioma | Astrocytoma |
| **Pineal region** | **Cerebellopontine angle** |
| Astrocytoma | Acoustic neuroma |
| Germ cell tumor | Epidermoid cyst |
| Pineoblastoma | Meningioma |
| Pineocytoma | **Spinal cord (epidural)** |
| **Cerebellum** | Metastatic carcinoma |
| Astrocytoma | **Spinal cord (extramedullary)** |
| Hemangioblastoma | Meningioma |
| Medulloblastoma | Neurilemoma |
| Metastatic carcinoma | Neurofibroma |
| **Third ventricle** | **Spinal cord (intramedullary)** |
| Choroid plexus papilloma | Astrocytoma |
| Colloid cyst | Ependymoma |
| Ependymoma | |

      b. **Osmotic agents** (such as intravenous infusion of hypertonic solution of mannitol) can very rapidly decrease intracranial pressure

   5. **Seizures** — tumors involving cerebral cortex can result in focal or generalized seizures

6. Spinal involvement — symptoms depend on level of tumor and whether extramedullary or intramedullary

   a. **Extramedullary tumors** — growth **outside spinal cord parenchyma** produces symptoms related to nerve root compression and bone destruction before spinal cord symptoms

   b. **Intramedullary tumors** — because of small size of spinal cord, **tumor growth within cord parenchyma** presents early as disturbance of spinal cord function (disruption of long tracts and local segmental signs)

B. General effects of tumors — from **increased intracranial pressure**

   1. Rigidity of cranium (following fontanelle and suture closure in childhood) allows no room for expansion; thus, increased pressure from tumor mass and surrounding brain edema is transmitted throughout ventricular system and brain

      a. **Obstructive hydrocephalus** — tumor mass and edema can distort ventricular system enough to **obstruct foramina** resulting in hydrocephalus

      b. **Herniation** — structures are displaced and distorted by edema and tumor mass resulting in protrusion (**herniation**) of portions of brain across midline (**transfalcial herniation** of cingulate gyrus), through tentorial incisura (**transtentorial herniation** of temporal lobe uncus), or through foramen magnum (**cerebellar tonsillar herniation**)

   2. **Papilledema** — **swelling of optic nerve head** (optic disc) results from increased cerebrospinal fluid pressure in subarachnoid space surrounding optic nerve; does not usually produce visual symptoms (can be associated with diplopia if abducens nerve palsy also occurs secondary to increased intracranial pressure)

   3. **Headache** — increased intracranial pressure produces intermittent headache; usually bilateral constant (not throbbing) headache, which is localized frontally and occipitally, increased with coughing or straining, and greatest in morning or after lying down

   4. Sudden exacerbation of signs and symptoms usually results from hemorrhage within tumor or acute herniation

   5. Death generally is from brain stem compression due to transtentorial herniation of medial temporal lobe or herniation of cerebellar tonsils through foramen magnum

II. **Diagnosis** of brain tumor

   A. Radiologic imaging studies (CT scan or MRI) — identify tumor mass

B. **Cerebral angiography** — identifies vascular supply to tumor prior to surgical approach

C. **Lumbar puncture** — identifies **malignant cells** in cerebrospinal fluid (CSF) in **meningeal carcinomatosis** or **leukemia**; lumbar puncture can result in **fatal herniation** with tumor mass

III. **Glioma**

   A. **Astrocytoma**

   1. **Diffusely infiltrative** tumor of **malignant astrocytes**; invades widely throughout brain, but does not metastasize outside central nervous system

      a. Adults — located in cerebral hemispheres; **most common primary malignant brain tumor**; can also occur as primary tumor of spinal cord

      b. Children — located in brain stem, optic nerves, and cerebellum

   2. Classification (grading) based on histologic appearance

      a. **Anaplastic astrocytoma** (WHO Grade 2 or 3) — malignant glioma **without necrosis**; survival approaches 2 years

      b. **Glioblastoma multiforme** (WHO Grade 4) — malignant glioma with histologic features of **necrosis**; survival 6-12 months

   3. Prognosis is generally poor

      a. In adults, treatment involves surgical debulking followed by radiation therapy

      b. In children, brain stem gliomas (usually of pons) identified by radiologic imaging studies are often treated with radiation therapy without surgical debulking

      c. Astrocytomas of cerebellum or optic nerve in children and of spinal cord in adults are low-grade and slow growing, resulting in extended survival

         (1) Cerebellar astrocytomas may be cystic with mural nodule that if totally resected can (rarely) result in surgical cure

         (2) **Optic gliomas** often occur as part of **neurofibromatosis**

   B. **Oligodendroglioma**

   1. Rare, slow-growing, **infiltrative** neoplasm of **malignant oligodendrocytes**, usually located in cerebral hemisphere (commonly **temporal lobe**)

2. Often associated with history of **chronic seizure disorder** that becomes progressively more refractory to anticonvulsant medications

3. Calcification present in most tumors is visible on radiologic imaging studies

4. Histologic study usually shows mixture of malignant oligodendrocytes and astrocytes; prognosis based on extent of astrocytic component, with lower percentage of astrocytes correlated with longer survival

5. Treatment involves extensive surgical debulking followed by radiation therapy

C. **Ependymoma**

1. Slow-growing tumor of malignant **ependymal cells**; common in childhood, usually arising in fourth ventricle

2. Treatment involves surgical debulking to relieve ventricular obstruction followed by radiation therapy; prognosis relates to degree of completeness of surgical resection, with long survival in some cases

IV. **Medulloblastoma (primitive neuroectodermal tumor; PNET)**

A. Common malignant brain tumor of childhood arising from primitive germinal cells in cerebellum

B. Usually arises in **midline of cerebellum**; occupies vermis and fourth ventricle with resultant hydrocephalus (from ventricular obstruction); produces signs of increased intracranial pressure (headache, vomiting, papilledema, and lethargy)

C. Rapidly growing tumor that spreads along cerebrospinal fluid (CSF) pathways to **metastasize throughout neuraxis**

D. Treatment involves surgical debulking and opening of cerebrospinal fluid pathways (to relieve hydrocephalus), followed by radiation therapy (tumor is very radiosensitive) to whole neuraxis (brain and spine) plus chemotherapy; aggressive therapy has produced many long-term survivors

V. **Retinoblastoma** — malignant tumor of **retina**

A. Common tumor of childhood presenting as **strabismus** (eye deviation), **leukokoria** ("cat's eye reflex" — **white reflex** due to mass in pupillary area behind lens)

B. Often **multifocal** within one eye or **bilateral**; when bilateral and in association with histologically-similar pineal tumor (**pineoblastoma**) has been termed **"trilateral retinoblastoma"**

C. **Familial** in 5% of cases; prototype tumor for type of cancer caused by mutational loss of genetic information

1. **"Two-hit" model of tumorigenesis** — retinoblastoma arises as result of two mutational events involving region on long arm of **chromosome 13** (region **13q14**)

    a. **No tumor** develops if both chromosomes of pair have normal region 13q14 or if only one chromosome of pair has deletion of 13q14 region (**heterozygous for deletion**)

    b. **Tumor develops** if both of chromosome 13 pair has deletion of 13q14 region (**homozygous for deletion**)

2. In hereditary (**autosomal dominant**) form of disease, abnormal chromosome 13 with deleted 13q14 region is inherited from one parent; subsequently, mutation occurs with 90% probability in 13q14 region of other chromosome 13 in some cells during development of eye, resulting in tumor growth from these abnormal cell clones

3. In sporadic form of disease, mutational events resulting in deletion of 13q14 region in both of chromosome 13 pair in same cell occur at sometime during eye development resulting in tumor cell clone with resultant tumor growth

D. Treatment involves **enucleation**; if tumor has not spread into optic nerve cure is possible; however, familial cases have predisposition to **second malignancies** (particularly **osteosarcoma of femur**)

VI. **Pineal region tumors**

A. Presenting symptoms

1. Obstruction of entrance to aqueduct of Sylvius results in hydrocephalus

2. Compression of midbrain tectum disturbs vertical gaze resulting in **inability to gaze upward (Parinaud's syndrome)**

B. Prognosis depends on histologic type of tumor

1. **Germ cell tumors** — malignant tumors with varying degrees of differentiation

    a. Good prognosis — **dysgerminoma** (similar to seminoma of testes and **highly radiosensitive**)

b. Poor prognosis — **choriocarcinoma** (with detectable blood and cerebrospinal fluid levels of **human chorionic gonadotropin**) or **embryonal carcinoma** (with detectable blood and cerebrospinal fluid levels of α-**fetoprotein**)

2. Malignant pineal parenchymal cell tumors (**pineoblastoma** or **pineocytoma**) have prognosis similar to medulloblastoma

3. **Astrocytoma** has prognosis similar to astrocytomas of cerebral hemispheres

VII. **Pituitary adenoma**

A. Benign tumors

1. Clinical division based upon radiologic imaging size into microadenoma (< 10 mm in diameter) or macroadenoma

2. Endocrinologic division into nonfunctioning (null cell) or hypersecreting (producing pituitary hormones)

B. Clinical presentations

1. Visual disturbance — impingement on optic chiasm from below results in **bitemporal hemianopsia** (initially, upper bitemporal quadrants more affected)

2. Endocrine disturbances

a. **Hyperprolactinemia**

(1) Common presentation is of young woman taking oral contraceptives who upon stopping medication is amenorrheic and has galactorrhea with serum prolactin levels > 100 ng/mL (normal < 15 ng/mL)

(2) Males experience **impotence, loss of libido**, and occasionally **gynecomastia**

(3) Must be differentiated from other causes of hyperprolactinemia (serum levels usually 15-100 ng/mL) including hypothalamic lesions and effect of medications (such as neuroleptics, reserpine, benzodiazepines, isoniazid)

b. Overproduction of **growth hormone**

(1) Before puberty — **gigantism**

(2) After puberty — **acromegaly**

c. **Cushing's disease**

(1) Truncal obesity, hypertension, muscle weakness, hirsutism, abdominal striae, glycosuria, osteoporosis, and personality changes (sometimes psychosis) resulting from **elevated levels of adrenocorticotrophic hormone (ACTH)** with consequent hypercorticism

(2) Must be differentiated from **Cushing's syndrome** due to excess glucocorticoid hormones alone, which can be produced by adrenal cortical hyperplasia (or adenoma) or result from exogenous corticosteroid administration (cortisone treatment)

C. Treatment

1. Surgical resection (transcranial or transsphenoidal) followed by radiation therapy if residual tumor remains

2. Since **dopamine** (released by hypothalamus) is normal inhibitor of prolactin release from pituitary cells, bromocriptine (dopamine receptor agonist) has been used to reduce prolactin levels and shrink prolactin-secreting tumors

3. Postoperative hormone replacement may be necessary if hormonal deficiency develops

D. Must be differentiated from:

1. **Pituitary apoplexy** — acute severe headache, ophthalmoplegia, loss of vision, subarachnoid hemorrhage, and lethargy or coma due to **hemorrhagic infarction** of pituitary adenoma; may be life threatening and requires immediate treatment with high-dose intravenous corticosteroids and surgical decompression

2. **Empty sella syndrome** — asymptomatic patient with enlarged sella turcica discovered incidentally in skull radiographs; subarachnoid space extends into sella turcica through an incompetent diaphragma sellae compressing pituitary gland and causing bony enlargement

3. **Craniopharyngioma** — slowly-growing **cystic tumor** with areas of calcification, arising from pituitary region **squamous cell rests** derived from **Rathke's pouch**; symptoms include visual field defects (from compression of optic chiasm), endocrine abnormalities (from compression of hypothalamus and pituitary), and headache and signs of increased intracranial pressure (from compression of third ventricle); treatment involves surgical resection

# CENTRAL NERVOUS SYSTEM NEOPLASMS

VIII. **Colloid cyst of third ventricle**

   A. Benign, epithelial-lined, gelatinous-filled, cystic tumor arising in **anterior roof of third ventricle**

   B. Produces **intermittent obstruction** ("ball valve") to cerebrospinal fluid flow through foramina of Monro

   C. Clinical symptoms of intermittent severe bifrontal **headache that changes with head position** (worse when lying on back, improved by sitting upright or bending forward); may present with acute coma leading to death (often investigated by medical examiner, with autopsy finding of third ventricular tumor)

   D. For tumors demonstrated during life by radiologic imaging studies, urgent ventricular decompression and neurosurgical resection of tumor is required

IX. **Meningioma**

   A. **Slow-growing, benign tumor** arising from leptomeningeal arachnoidal cells; usually tightly bound to dura

   B. Relatively common, **more frequent in women** than men, and often found incidentally at autopsy

   C. Central nervous system injury by compression; because of slow growth may become very large before significant clinical symptoms develop

   D. Treatment is complete surgical resection

X. **Schwannoma (Neurilemoma)**

   A. Slow-growing, encapsulated benign **tumor of peripheral or cranial nerve** arising from Schwann cells

   B. Most common **primary intraspinal tumor**; commonest intracranial site is **acoustic nerve** ("**acoustic neuroma**"); can also occur along **peripheral nerves**

   C. Can be part of **neurofibromatosis (von Recklinghausen's syndrome)** — autosomal dominant **neurocutaneous syndrome (phakomatosis)**

   1. **Neurofibromatosis type 1** (NF-1, peripheral or classical neurofibromatosis) — multiple **café-au-lait spots**, axillary freckling, **Lisch nodules** (iris hamartomas appearing as small yellow or brown elevations), cutaneous neurofibromas, acoustic neuromas (schwannomas), optic gliomas, spinal nerve root neurofibromas or schwannomas (neurilemomas), and

skeletal anomalies; gene localized to chromosome 17 involving protein **neurofibromin** (GTPase activating protein)

    2. **Neurofibromatosis type 2** (NF-2, **central neurofibromatosis**) — **bilateral acoustic neuromas** (schwannomas); multiple meningiomas; rare café-au-lait spots; gene localized to chromosome 22 involving protein **merlin** (cytoskeleton-associated protein)

  D. Treatment involves complete surgical resection

XI. **Hemangioblastoma** of cerebellum

  A. Slow-growing, cystic, highly vascular, benign tumor arising from capillary endothelium; often associated with **polycythemia** due to erythropoietic factor elaborated by tumor

  B. Clinical presentation of ataxia, vertigo or dizziness, headache, and occasionally papilledema and other signs of increased intracranial pressure

  C. About 10% of cases are part of **Von Hippel-Lindau syndrome** (cerebellar hemangioblastoma, retinal angiomatosis, renal and pancreatic cysts, and renal cell carcinoma), which is associated with gene defect on short arm of chromosome 3

  D. Treatment involves complete surgical resection

XII. **Primary central nervous system lymphoma**

  A. Occurs in individuals with **immune compromise** (as in **AIDS**) or in older individuals (generally over age 60 years) with no evidence of immunologic disorder

  B. **Deep cerebral hemisphere** tumor, often **bilateral** and showing homogenous contrast enhancement on radiologic imaging studies

  C. Histologically **high grade** lymphoma, often **immunoblastic (B-cell)** type

  D. Initial dramatic response to **corticosteroids** and **radiation** ("disappearing tumor") usually followed by recurrence and fatality within 3 years

XIII. Metastatic tumors

  A. **Parenchymal brain** metastases

    1. Most common central nervous system **tumor**; may be single or multiple; affects almost 20% of patients dying of cancer

2. Any malignancy can potentially metastasize to central nervous system, but most common are **lung cancer**, **breast cancer**, and **malignant melanoma**

3. Frequency of metastatic site is roughly proportional to size of brain region (cerebrum > cerebellum > brain stem)

4. Since metastatic tumors tend to push into brain, rather than diffusely infiltrate (like gliomas), there is often seemingly well-defined margin; however, there is **extensive edema** in surrounding brain

5. Clinical presentation includes **headache, papilledema**, and signs of increased intracranial pressure; focal signs depend upon specific site of nervous system involvement

6. Treatment depends on location and number of metastases, type of primary tumor, degree of systemic involvement, and presumed general prognosis

    a. **High-dose corticosteroids** reduce edema and can dramatically improve symptoms (particularly headache and signs of increased intracranial pressure)

    b. **Surgical resection** indicated for solitary, readily accessible brain lesion or lesion involving spinal cord

    c. **Radiation therapy** indicated for multiple metastases or when patient is not surgical candidate because of systemic condition

B. **Meningeal carcinomatosis, lymphoma, or leukemia**

1. Common **complication of cancer**, particularly lymphoma, leukemia, small-cell ("oat-cell") lung carcinoma, breast carcinoma, "signet-ring cell" gastric carcinoma, and choriocarcinoma

2. Clinical presentation is of signs of **meningeal irritation** (backache, headache, and stiff neck) and focal infiltration of cranial nerves or spinal nerve roots (cranial nerve palsies or radicular pain and weakness)

3. **Cerebrospinal fluid examination** reveals malignant cells, along with inflammatory cells (lymphocytes and neutrophils), elevated protein level, and decreased glucose level (carcinomatous meningitis)

4. Treatment involves radiation therapy to whole neuraxis (craniospinal radiation) and injection of chemotherapeutic drugs directly into subarachnoid space (intrathecal chemotherapy); prognosis is poor since meningeal involvement usually indicates widespread metastatic disease

C. **Spinal metastases**

1. Usually result from **local spread to epidural space** by direct extension out of involved vertebral body or through intervertebral foramina; most common in thoracic and lumbar vertebrae

2. Initial symptoms include **unremitting** severe localized back or radicular **pain**, weakness or numbness in legs, and urinary retention; rapidly progresses to complete **paraplegia** (from vascular compression with consequent spinal cord infarction)

3. Treatment involves administration of **high-dose corticosteroids** to reduce edema and surgical decompression or radiotherapy to reduce tumor bulk and relieve pressure on spinal cord

# Chapter 12 — MOVEMENT DISORDERS

I. Diseases of the motor system have often been loosely divided into those of pyramidal system and of extrapyramidal system

   A. **Pyramidal motor system** — upper motor neuron originating in cerebral motor cortex of precentral gyrus and its axonal extensions traveling through internal capsule, cerebral peduncle, basis pontis, medullary pyramid, and spinal cord lateral column to directly control lower motor neurons of cranial nerve motor nuclei and spinal cord anterior horns

   B. **Extrapyramidal motor system** — neurons and their axonal processes originating in deep cerebral nuclei (basal ganglia), brain stem, and cerebellum which influence motor function by modulating pyramidal motor system

II. Features of abnormal motor activity

   A. **Tremor** — involuntary, **rhythmic oscillations** around joint

   1. **Physiologic tremor** — **normal** fine 10 Hz **tremor in all muscle groups,** usually only barely visible

   2. **Pathologic tremor** — clinically apparent abnormal tremor

      Parkinsonian resting tremor.

      a. **Resting (parkinsonian) tremor** — coarse 3-5 Hz tremor **present when the limb is held still,** but **suppressed by voluntary movement** (tremor resuming when limb again comes to rest)

      b. **Action (postural) tremor** — regular or irregular tremor of varying frequencies present whenever muscles are **activated for movement** and absent at rest; most common form is **benign essential tremor** (familial tremor or senile tremor) which is autosomal dominantly inherited, presents from adolescence to late adult years, and is reduced by alcohol or propranolol (Inderal)

      c. **Intention (ataxic, cerebellar) tremor** — irregular 2-4 Hz tremor that **interrupts limb movement** (greatest as endpoint of movement is approached); **absent at rest;** usually associated with other signs of cerebellar dysfunction such as ataxia, hypotonia, and trunk and head oscillations (titubation)

B. **Myoclonus** — **sudden, irregular, shock-like contractions** of one or more muscles; associated with epilepsy (**myoclonus epilepsy**) or in recovery from hypoxic encephalopathy (intentional or action myoclonus)

C. **Tics (habit spasms)** — idiosyncratic, often complex and **stereotyped movements and mannerisms**; usually correspond to useful purposeful acts or movements, but performed repeatedly without apparent purpose; most common tic disorder is **Gilles de la Tourette syndrome**

D. **Choreoathetosis** — chorea and athetosis are distinct movement disorders, but often occur together and individual movements blend into one another; observed in **Huntington's disease**, **Sydenham's chorea** (associated with rheumatic fever and ß-hemolytic streptococcal infection), **Wilson's disease**, and **athetoid cerebral palsy** (**kernicterus** or perinatal hypoxic-ischemic encephalopathy)

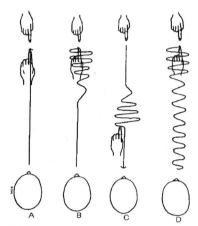

On finger-to-nose test, normals point smoothly and accurately (A); amplitude of intention tremor increases near target (B, C); action tremor has same amplitude throughout range of motion (D).

   1. **Chorea** — **purposeless, abrupt, rapid, irregular, involuntary, nonrepetitive, jerky movements**; to make movements less noticeable, patient may attempt to incorporate choreic jerks into voluntary movement, but such superimposition makes voluntary movement appear exaggerated and awkward

   2. **Athetosis** — **purposeless, slow, sinuous, writhing movements** which appear to flow into one another and interrupt attempts to maintain posture or to initiate voluntary activity

E. **Dystonia (torsion spasm)** — **sustained abnormal or inappropriate extreme posture**; positioning resembles an extreme form of athetosis in which the change of position is exceptionally slow; observed in **dystonia musculorum deformans** (hereditary progressive dystonia), spasmodic torticollis, choreoathetotic diseases, and as acute reaction to dopamine-blocking neuroleptic drugs (such as haloperidol or chlorpromazine) particularly involving eye muscles (**oculogyric crisis**)

Dystonic posture of right hand.

F. **Ballismus** — purposeless, violent, flinging involuntary extremity movement, usually unilateral (hemiballismus) involving arm; results from lesion (infarct or hemorrhage) of contralateral **subthalamic nucleus of Luys**

G. **Asterixis** — frequent (several times per minute) arrhythmic lapses of posture associated with brief loss of electromyographic activity (muscle silence); called *"liver flap"* because of frequent association with **hepatic encephalopathy**

H. **Tardive dyskinesia** — involuntary movements that develop following continuous use of **dopamine-blocking neuroleptic drugs** (related to duration, dose, and drug potency)

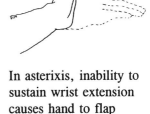

In asterixis, inability to sustain wrist extension causes hand to flap (bye-bye gesture).

   1. Repetitive purposeless movements (resembling choreoathetosis) usually restricted to head and neck, particularly oral-buccal-lingual musculature (tongue protrusion, chewing, lip smacking, grimacing), but may be more generalized

   2. Results from progressive medication-induced alteration of brain post-synaptic dopamine receptors; cessation of medication may transiently increase abnormal movements which then slowly resolve over months to years

I. **Clonus** — rhythmic muscle contractions precipitated by sudden muscle stretch associated with other signs of **spasticity**; due to damage to **pyramidal motor system** (upper motor neuron); most commonly observed at ankle

J. **Rigidity** — **increased muscle tone** present **throughout range of motion** of limb, with constant and uniform resistance to passive movement like that noted in attempting to bend lead pipe; observed in parkinsonism and diseases of basal ganglia

K. **Spasticity** — **increased muscle tone** and **increased tendon reflexes** associated with damage to **pyramidal motor system**; during rapid passive limb movement, there is abrupt "catch" and increasing resistance to movement followed by abrupt reduction in resistance as movement continues ("clasp knife" phenomenon)

L. **Focal motor seizures** — rhythmic movements related to electrical seizure activity originating in cerebral motor cortex

III. **Parkinson's syndrome (parkinsonism)**

A. **Triad** of **resting tremor, rigidity,** and **bradykinesia**, along with postural, gait and handwriting disturbances

   1. **Resting tremor** — present when limb is still; in hands has "pill rolling" quality

   2. **Rigidity** — increased tone throughout the range of movement; during passive movement, ratchet quality (**cogwheel phenomenon**) due to tremor may be evident

3. **Bradykinesia** — slowed or lack of movement; relative immobility with little spontaneous movement; "masked face" is due to relatively **immobile, expressionless** facial appearance

4. **Postural changes** include internally-rotated, stooped shoulders along with neck, back, and knee flexion; gait problems include difficulty initiating movement and initial slow, small steps (marche à petit pas) which then increase in speed such that patient seems to be falling forward and have difficulty stopping (**festination**); trunk movements are not fluid but stiff (en bloc); handwriting is slow and small with letters tightly bunched (**micrographia**)

B. Most common cause of Parkinson's syndrome is pharmacological due to administration of drugs (such as **neuroleptics**) with **dopamine antagonist** properties; cessation of such medications should result in reversal of symptoms

C. **Parkinson's disease** - idiopathic progressive disorder, primarily occurring in middle or late adult life, in which neurons in **midbrain substantia nigra** (and to lesser extent **pontine locus ceruleus**) degenerate, accompanied by **Lewy body inclusions** in remaining neurons

D. Pharmacologic treatment includes:

1. **Anticholinergic agents** such as benztropine mesylate (Cogentin) or trihexyphenidyl (Artane)

2. **Amantadine hydrochloride** (Symmetrel), which is antiviral agent, but also augments dopamine release from presynaptic terminals

3. **L-dopa** in combination with peripheral dopa decarboxylase inhibitor carbidopa (Sinemet); L-dopa is converted to dopamine in residual substantia nigra neurons; may result in "on-off" phenomenon, in which patient suddenly develops transient hyperkinesia (choreic movements) followed by return to immobility

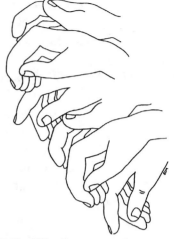

"Pill-rolling" tremor of parkinsonism.

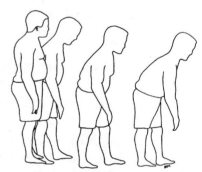

Typical parkinsonian posture and gait, which becomes faster (festination).

Parkinsonian handwriting abnormalities include micrographia.

# MOVEMENT DISORDERS

4. **Bromocriptine** (Parlodel) or pergolide (Permax) directly stimulate dopamine receptors

5. **Deprenyl** (Selegiline), inhibitor of **monoamine oxidase-B** (MAO-B), prevents dopamine breakdown and possibly also protects against neurotoxins that damage substantia nigra neurons

E. Parkinsonism associated with autonomic dysfunction (including orthostatic hypotension and impotence) is **Shy-Drager syndrome**; parkinsonism associated with loss of voluntary eye movements (particularly upgaze) is **Steele-Richardson-Olszewski progressive supranuclear palsy**; destruction of substantia nigra neurons by meperidine analog designer drug **MPTP** (1-methyl-4-phenyl-1,2,3,6-tetrahydropyridine) also results in parkinsonism

IV. **Huntington's disease (Huntington's chorea)**

A. Degenerative disorder with **progressive choreoathetosis and dementia**; psychiatric symptoms and suicide are common during early phase of disease

B. Average onset is age 35 years; onset before age 15 years (juvenile Huntington's disease) is characterized by rigidity and seizures

C. **Dominantly-inherited disorder** linked to abnormal gene on short arm of chromosome 4

1. Genetic abnormality consists of expansion (increased length) of unstable trinucleotide (CAG) repeat sequence in gene

2. **Anticipation** — increasing severity of symptoms over generations; relates to increasing length of trinucleotide repeat sequence which correlates with earlier onset of symptoms

D. CT scan and MRI show **atrophy of caudate nucleus** with consequent ventricular enlargement

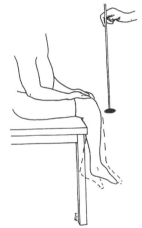

Excessive number of swings with pendular reflexes.

V. **Wilson's disease (hepatolenticular degeneration)**

A. Progressive disorder characterized by **ataxia, choreoathetosis, dystonia**, and **dysarthria**, along with **hepatic dysfunction** (cirrhosis)

B. Copper deposits at corneal limbus (**Kayser-Fleischer rings**) identifiable by slit-lamp exam

C. Autosomal recessive disorder (gene locus on long arm of chromosome 13) resulting in abnormal copper metabolism; laboratory studies reveal **low serum ceruloplasmin, high urinary copper excretion**, and elevated levels of copper in liver

D. Treatment consists of copper chelation using D-penicillamine and low copper diet

VI. **Cerebellar ataxia**

A. Movement disturbance **ipsilateral** to side of **cerebellar hemisphere dysfunction**

B. Examination shows difficulty with rapid alternating movements, hypotonia, pendular reflexes, and abnormal rebound

C. Intention tremor evident on finger-to-nose and heel-knee-shin tests

D. Loss of normal speech melody with explosive quality (ataxic dysarthria or scanning speech)

Abnormal rebound: sudden release of resistance will cause patient's arm to strike face.

VII. **Gilles de la Tourette syndrome**

A. Syndrome of **progressive multiple tics** involving mainly face, head, and shoulders; associated respiratory tics, grunts, and snorts eventually develop into **vocal tics** consisting of **repeated obscenities (coprolalia)**

B. Onset usually before adolescence and more common in males

C. Pharmacologic treatment includes dopamine-blocking drugs haloperidol and pimozide (Orap) and $\alpha_2$-adrenergic receptor agonist clonidine (Catapres)

# Chapter 13                                                   CENTRAL NERVOUS SYSTEM INFECTIONS

I. Central nervous system infections are usually **life-threatening**, necessitating immediate diagnosis and treatment to prevent death or significant residual brain damage

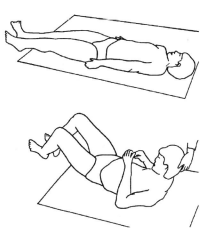

Brudzinski's sign: passive neck flexion causes flexion of thighs and legs.

    A. **Headache** with **fever** and signs of **meningeal irritation (nuchal rigidity** and **Brudzinski's sign)** strongly suggests infection which must be ruled out without delay; however, neonates or elderly individuals with central nervous system infection may not be able to report headache and fever, and signs of meningeal irritation may be absent

    B. Predisposing factors for central nervous system infections include **bacteremia** (such as in **bacterial endocarditis**), debilitating conditions (such as **chronic renal failure**), **immunologic compromise** (such as from **AIDS, lymphoma**, or immunosuppressive drugs), disruption of protective barriers (such as following **basilar skull fracture** with associated cerebrospinal fluid **rhinorrhea** or following neurosurgical procedures such as cerebrospinal fluid shunt placement)

II. **Cerebrospinal fluid (CSF) examination**

    A. **Most important test for identifying evidence of central nervous system infection is** *examination of cerebrospinal fluid*

    B. **Cerebrospinal fluid** study provides definitive evidence of infection along with opportunity for culture and identification of infecting organism; must be performed *prior* to instituting antibiotic therapy in order to culture most organisms; culture of material obtained from sources other than cerebrospinal fluid (such as blood culture, urine culture, or wound culture) will usually either not yield organisms or yield organisms other than that producing central nervous system infection

    C. **Lumbar puncture** (performed by attending physician) is preferred method for obtaining cerebrospinal fluid in most cases; however, cerebrospinal fluid can also be obtained by **cisterna magna or lateral cervical puncture** (performed by neurosurgeons under fluoroscopy) or by **ventricular puncture** (performed by neurosurgeons through cranial burr hole, except when cerebrospinal fluid shunt tubing is already in place, or through open fontanelle in neonates)

D.  Caution must be exercised in performing lumbar puncture in patients with signs of **increased intracranial pressure** (such as **papilledema**), since removal of cerebrospinal fluid from subarachnoid space may alter intracranial pressure dynamics precipitating brain **herniation** (transtentorial uncal herniation, central rostral-caudal herniation, or cerebellar tonsillar herniation)

   1.  Since delay in diagnosis and treatment of central nervous system infection is usually fatal, attending physician must balance potential risks of lumbar puncture versus potential risk of delay in diagnosis

   2.  **Radiologic imaging studies** (such as CT scans) can quickly identify factors predisposing to herniation (such as mass effect or pre-existing early herniation)

   3.  Inserting secure intravenous infusion apparatus prior to performing lumbar puncture permits infusion of hyperosmolar agents (such as mannitol) to reduce intracranial pressure if signs of herniation appear during lumbar puncture; additionally, antibiotic therapy can be immediately instituted once appropriate cerebrospinal fluid samples have been obtained

E.  Lumbar puncture should be performed using meticulous **sterile technique**, obtaining values for **opening pressure** and removing sufficient cerebrospinal fluid for all necessary studies

   1.  Normal **opening pressure** should be no more than **180 mm of water**

   2.  It is incorrect that removing only minimal (1 mL) cerebrospinal fluid will prevent herniation; once arachnoid membrane is punctured, changes in pressure dynamics will occur, since after removal of spinal needle, at least 40 mL or more of cerebrospinal fluid will continue to leak through puncture holes in arachnoid and dura (filling subdural and epidural spaces)

F.  Normal cerebrospinal fluid is **thin, colorless, sparkling, crystal-clear fluid** that does not coagulate; abnormal cerebrospinal fluid can be **turbid** (cloudy), colored, viscous, frankly purulent, or bloody; as few as 400 blood cells per cubic millimeter (microliter) result in turbidity

G.  Tests performed on cerebrospinal fluid include:

   1.  **Glucose** — normal cerebrospinal fluid glucose level is about **60% of blood glucose level**; values of **less than 50%** of simultaneous blood glucose or values **below 40 mg/dL** are indicative of meningeal inflammation (as occurs with central nervous system infection); very low glucose values (less than 20 mg/dL) are indicative of granulomatous infections such as tuberculous meningitis

2. **Protein** — cerebrospinal fluid protein varies with central nervous system site and with age; protein elevation is found with variety of central nervous system disorders including inflammation, infection, tumor, or stroke

3. **Cell count** — cerebrospinal fluid normally contains *no neutrophils* or polymorphonuclear leukocytes and *no more than five lymphocytes* per cubic millimeter (microliter); more cells (cerebrospinal fluid **pleocytosis**) are abnormal: neutrophils are associated with infection, hemorrhage, infarction (stroke), and tumor; eosinophils occur in small numbers in same circumstances as neutrophils, while large numbers are found with central nervous system parasitic infections, foreign bodies, allergic processes, malignant lymphomas, and post-myelography; plasma cells are found in viral diseases, multiple sclerosis, in recovery phase of bacterial meningitis, and in luetic (syphilis) infection

4. Microbiology

    a. Light microscopy — if cerebrospinal fluid contains more than 100 organisms per milliliter, microscopic slides made from sediment obtained by centrifugation will reveal bacteria or fungi after **Gram stain**; tuberculous bacilli can be visualized in slides stained with **acid-fast** methods (Ziehl-Neelsen or fluorescent rhodamine stains); identification of *Cryptococcus neoformans* is possible in **India ink preparations**; amoeba can be detected as mobile trophozoites in wet mounts examined by phase interference microscopy; neutrophils containing keratin fragments can be identified using polarized light in chemical meningitis secondary to spillage of contents of epidermoid tumor, craniopharyngioma, or dermoid cyst

    | CSF PROTEIN (upper limits) | |
    | --- | --- |
    | **Young adult by site** | |
    | Ventricular | 15 mg/dL |
    | Cisterna magna | 25 mg/dL |
    | Lumbar | 45 mg/dL |
    | **Lumbar fluid by age** | |
    | < 30 days | 150 mg/dL |
    | 30 days - 3 months | 100 mg/dL |
    | 3 months - 6 months | 50 mg/dL |
    | 6 months - 10 years | 30 mg/dL |
    | 10 years - 40 years | 45 mg/dL |
    | 40 years - 50 years | 50 mg/dL |
    | 50 years - 60 years | 55 mg/dL |
    | > 60 years | 60 mg/dL |

    b. Serologic studies (antigen-antibody studies)

        (1) **Cryptococcal antigen** (latex particle agglutination test)

        (2) Elevated IgM levels can be detected in herpes simplex encephalitis

        (3) **Countercurrent immunoelectrophoresis (CIE)** can rapidly detect polysaccharide antigens associated with meningococcal, pneumococcal, and *Haemophilus influenzae* (type B) meningitis

(4) **Syphilis serology** — **VDRL** procedure can identify up to 50% and fluorescent treponemal antibody absorption test (**FTA-ABS**) will identify over 90% of patients with neurosyphilis

(5) IgG and measles antibody titers are elevated in subacute sclerosing panencephalitis

c. Culture

(1) Final identification and verification of organism type requires **culture**; however, contamination can lead to false positive results and negative culture results do not exclude organisms, since technical problems may prevent culture; false negative cultures result from antimicrobial drug treatment prior to cerebrospinal fluid collection (**partially-treated meningitis**)

(2) Culture of some organisms requires special media or collection techniques (for example, large quantities of cerebrospinal fluid are necessary for successful culturing of tuberculous bacilli and special transport media are necessary to preserve many viruses)

III. **Bacterial (suppurative; purulent) meningitis**

A. Organisms typically causing bacterial meningitis vary with age:

1. **Neonates** (up to age 3 months)

a. Lack of sufficient immunologic maturity to mount adequate immune response allows local infections to become disseminated to central nervous system; presenting signs and symptoms are non-specific (such as **irritability, poor feeding, cyanosis, jitteriness**, or **lethargy**); nuchal rigidity does not occur; fever is uncommon, and infected infant often is hypothermic

b. Infection by organisms of female genital tract is acquired during passage through birth canal; such organisms include: **enteric gram-negative rods** (particularly *Escherichia coli*), **group B ß-hemolytic streptococci**, and *Listeria monocytogenes*; such infections usually present during first week of life

c. Infections with *Staphylococcus aureus, Pseudomonas, Proteus*, and *Salmonella* are usually acquired after birth due to environmental contamination (contact with contaminated articles or handling by infected individuals)

# CENTRAL NERVOUS SYSTEM INFECTIONS

2. **Infants**

   a. Immunologic system is mature enough by age 3 months to protect against most environmental pathogens, except against those organisms surrounded by **polysaccharide capsule:** *Haemophilus influenzae* (particularly group B) and *Streptococcus pneumoniae* **(pneumococcus)**; loss of protective maternal immunoglobulins (acquired through transplacental passage) against these organisms permits nasopharyngeal colonization to progress to localized infection (including otitis media), bacteremia, and in some infants, meningitis

   b. Peak incidence at age 1 year; after age 4 years meningitis is uncommon (due to further maturation of immune system)

3. **Children and adolescents**

   a. Because of large number of type specific capsular antigens and lack of cross immunity between types, *Streptococcus pneumoniae* **(pneumococcus)** remains common as cause of meningitis in all age groups after age 3 years

   b. *Neisseria meningitidis* **(meningococcus)** produces epidemic meningitis with rapid evolution of symptoms and associated petechial or purpuric skin rash

4. Adults

   a. Meningitis is rare and most often is associated with some underlying **predisposing condition** such as **immunologic compromise, alcoholism, trauma**, surgery, or systemic illness (such as pneumonia, septicemia, or endocarditis); **recurrent meningitis** (particularly pneumococcal) suggests **abnormal communication** between subarachnoid space and skin or nasopharynx; splenectomy or sickle cell anemia (with its associated auto-splenectomy) predisposes to pneumococcal sepsis and meningitis

   b. Elderly individuals seem to be particularly susceptible to meningitis caused by same organisms as in neonates or infants

B. Complications

   1. **Increased intracranial pressure**

      a. Obstruction of normal subarachnoid cerebrospinal fluid circulation due to inflammation produces increased pressure that resolves as inflammation subsides with treatment

      b. **Cerebral edema** may result from inflammation and bacterial toxins

c. **Cortical venous thrombophlebitis** or **arteritis** involving subarachnoid vessels can produce brain ischemia or **infarction** with resultant edema

2. **Syndrome of inappropriate antidiuretic hormone (ADH; vasopressin) secretion (cerebral salt wasting)**

    a. Derangement (due to subarachnoid inflammation or cerebral edema) of hypothalamic mechanisms regulating blood volume and osmolality resulting in **water retention** with consequent **dilutional hyponatremia**

    b. Persistent ADH secretion despite fall in serum osmolality below normal values; diagnosis based upon finding **hyponatremia** with concomitant **serum hypo-osmolality** and **urinary hyperosmolality**

    c. Hyponatremia is usually not associated with clinical symptoms until serum sodium level falls below 120 meq/L and seizures and coma may occur with values approaching 100 meq/L levels

    d. Treatment involves **fluid restriction**; treatment with combination hypertonic saline infusion and diuretics can rapidly correct serum sodium level but also causes **central pontine myelinolysis**

3. **Brain infarction**

    a. **Thrombophlebitis** — cortical veins encased in purulent exudate within subarachnoid space can become thrombosed resulting in venous infarction

    b. **Arteritis** — subarachnoid arteries encased in purulent exudate become inflamed (vasculitis) with consequent thrombosis and infarction

    c. **Abscess** — dead tissue in areas of infarction can become nidus for development of brain abscess

        (1) Identified by radiologic imaging studies

        (2) Must be differentiated from primary brain abscess which leaks organisms into subarachnoid space producing secondary meningitis

4. **Subdural fluid collection**

    a. **Subdural effusion — infants with meningitis** are particularly susceptible to development of subdural fluid collections; differentiated from subarachnoid (cerebrospinal) fluid by **high protein** content of subdural fluid; signs of subdural fluid collection include increased **irritability**, lethargy, **seizures**, **persistent fever**, fullness of fontanelle, or increasing head circumference; small effusions resolve spontaneously, but large effusions require repeated percutaneous subdural taps to remove fluid until it no longer accumulates

b. **Subdural empyema** — extremely rare complication of meningitis in which subdural effusion fluid becomes infected; treatment involves immediate neurosurgical drainage

5. **Seizures**

    a. Frequent in infants and children presumably related to effects of subarachnoid inflammation and bacterial toxins on adjacent cerebral cortex; can also occur with subdural effusion or empyema (local pressure effect) or with metabolic derangements (such as hyponatremia from syndrome of inappropriate ADH secretion); seizures usually stop with successful treatment of meningitis and do not lead to epilepsy

    b. Can also occur due to cortical infarction or brain abscess; subsequent epilepsy is likely due to permanent cerebral damage

6. **Persistent fever** — indicates **inadequate antibiotic therapy, subdural effusion** or empyema, **brain abscess**, or **systemic infectious focus** (such as abscess elsewhere in body or osteomyelitis)

7. **Hydrocephalus**

    a. Acute — purulent material can obstruct cerebrospinal fluid pathways leading to ventricular enlargement; diagnosed by radiologic imaging studies; usually resolves with treatment of infection

    b. **Communicating hydrocephalus** — long-term complication due to **subarachnoid fibrosis** as reparative reaction following meningitis; fibrosis obstructs normal cerebrospinal fluid flow toward arachnoid granulations (site of absorption into venous sinuses); diagnosed by radiologic imaging studies; necessitates cerebrospinal fluid shunt insertion after complete cure of infection

8. Persistent neurologic deficits

    a. **Cranial nerve palsies** — palsies related to increased intracranial pressure (involving oculomotor, abducens, and facial nerves) usually resolve with treatment of infection; persistent damage to **vestibuloacoustic nerve (cranial nerve VIII)**, particularly manifested as **hearing loss**, is common complication of meningitis in infants and children; visual disturbances are also common

    b. Infarction or abscess results in permanent brain destruction with consequent neurologic deficits

c. Other long-term sequelae in children include **mental retardation, behavioral disorders**, and **learning disabilities**

9. Recurrent meningitis — suggests communication between subarachnoid space and skin or mucous membranes, parameningeal focus of infection (such as cranial osteomyelitis), or inadequate antibiotic treatment

10. **Waterhouse-Friderichsen syndrome — fulminant meningococcemia** can produce septic shock and vasomotor collapse associated with **hemorrhagic infarction of adrenal glands**; often rapidly fatal

11. **Ventriculitis** — infection of intraventricular cerebrospinal fluid; uncommon except associated with abscess that ruptures into ventricle or with neurosurgical opening to ventricles (as with cerebrospinal shunt tubing); often rapidly fatal except when caused by low virulence organisms such as *Staphylococcus epidermidis* or enterococcus (group D streptococci)

C. Treatment

1. *Bacterial meningitis requires immediate treatment (medical emergency)*

2. *Antibiotic therapy must be administered intravenously and at high doses*; minimum treatment is for 10 days (or 7 days after becoming afebrile), but for relatively resistant organisms (such as gram-negative enteric bacteria) treatment for 3 weeks or longer often is necessary

3. Cerebrospinal fluid is examined from lumbar punctures performed 24-48 hours after initiating antibiotic therapy and 24-48 hours after cessation of antibiotics:

    a. Cerebrospinal fluid should be *sterile* by 48 hours after beginning therapy for relatively sensitive infecting organisms (such as *Neisseria meningitidis*, *Haemophilus influenzae*, or *Streptococcus pneumoniae*)

    b. Positive culture at 48 hours can sometimes still be obtained in relatively resistant organisms (such as gram-negative enteric bacteria), necessitating repeat lumbar puncture at 72-96 hours after antibiotic initiation, at which time cerebrospinal fluid should be sterile

    c. With successful antibiotic therapy, cerebrospinal fluid differential cell count changes to **predominantly lymphocytes**

    d. Persistence of organisms identifiable by culture or **persistent neutrophilic predominance** in cell count signifies **inadequate antibiotic therapy** or **parameningeal focus** with continual seeding of cerebrospinal fluid

4. Antibiotic choice

    a. Initial therapy (prior to culture results) should be based upon assumptions concerning likely organism based upon patient age, environmental exposure, and risk factors; combinations of antibiotics covering a wide-range of potential pathogens is preferred

        (1) Combination therapy with ampicillin/gentamicin or ampicillin/cefotaxime for neonates

        (2) Combination therapy with ampicillin/ceftriaxone or ampicillin/chloramphenicol for infants, children, adolescents, and adults

    b. Once culture results are available, therapy should be changed to antibiotic to which organism is sensitive

5. Prophylaxis

    a. All infants should receive **immunization** against *Haemophilus influenzae* type B

    b. Immunization against many serotypes of pneumococcus is available for elderly and immunocompromised individuals

    c. During epidemics of meningococcal meningitis, contacts should receive prophylactic antibiotics

IV. **Tuberculous (*Mycobacterium tuberculosis*) infection**

A. Since organism culture takes as long as four weeks, diagnosis must be based upon finding evidence of central nervous system infection (such as elevated cerebrospinal fluid cell count) along with **active pulmonary tuberculosis, positive purified protein derivative (PPD) skin test**, and microscopically identifiable acid-fast organisms in cerebrospinal fluid sediment or brain tissue biopsy

B. **Tuberculous meningitis**

1. Gradual onset of nonspecific symptoms of **malaise, low-grade fever**, and **anorexia** followed later by lethargy, seizures, **cranial nerve palsies**, focal (often brain stem) neurologic signs, and evidence of increased intracranial pressure (papilledema); meningeal signs (such as nuchal rigidity) are uncommon

2. Cerebrospinal fluid obtained by lumbar puncture usually shows increased numbers of **lymphocytes, low glucose** (often less than 20 mg/dL), and **elevated protein** (sometimes fluid will form clot or pellicle)

C. **Tuberculoma** — **granuloma** involving **brain parenchyma** (can be single or multiple); must be differentiated from tumor or abscess, necessitating neurosurgical biopsy with culture and histologic identification of tuberculous organisms

D. Treatment requires long course of combination antituberculous medications; initial medication regimen must be modified when culture results establish sensitivity

E. Complications

1. Medication-induced neurologic sequelae include:

   a. **Isoniazid-induced neuropathy** — can be avoided by **pyridoxine supplementation**

   b. Streptomycin-induced damage to cranial nerve VIII (with consequent **deafness** and **vestibular dysfunction**

   c. Ethambutol-induced optic nerve damage

   d. Cycloserine-induced seizures

2. **Hydrocephalus** — very common due to dense **subarachnoid fibrosis** surrounding brain stem and interfering with normal cerebrospinal fluid flow

3. **Arteritis** — involves vessels at base of brain, producing **infarctions**

V. **Brain abscess**

A. Slowly progressive onset of symptoms suggesting **intracranial mass lesion**, including **headache, focal neurologic signs, lethargy,** and **increased intracranial pressure**; **seizures** are frequent, but fever and leukocytosis often are absent; predisposing conditions include **sinusitis, mastoiditis**, bacterial **endocarditis**, or lung abscess

B. Diagnosis suggested by radiologic imaging studies (particularly **"doughnut" lesion** consisting of low density area with encircling higher density capsule and marked surrounding cerebral edema); confirmation requires neurosurgical exploration and drainage (with biopsy and **culture** to determine organism)

C. Lumbar puncture can be dangerous due to potential for inducing **herniation**, and cerebrospinal fluid study generally is not helpful (culture is usually negative, cell count normal, and protein only slightly elevated)

D. Evolution of brain abscess — initial phase is **cerebritis** (poorly marginated area of infection and edema); followed by **central necrosis** and liquifaction which is then walled off by margin of dense fibrosis and gliosis (evident as "doughnut" lesion on radiologic imaging studies)

## CENTRAL NERVOUS SYSTEM INFECTIONS

    E. Treatment involves neurosurgical **drainage** and high dose intravenous antibiotics

VI. **Fungal infection**

    A. Can present either as **chronic meningitis** (mimicking tuberculous meningitis) or as parenchymal infection (mimicking brain abscess)

    B. Most often affects **immunocompromised individuals** (although cryptococcal meningitis can occur in otherwise healthy individuals and candidal meningitis can affect neonates)

    C. Selected common pathogenic organisms

        1. *Cryptococcus neoformans* — symptoms of meningitis can be rapidly progressive and fatal or indolent over many months; large volumes of cerebrospinal fluid (40 mL or more) are often necessary to establish diagnosis by culture, microscopic examination (Gram stain and India ink preparation), or antigen assay

        2. *Candida albicans* — usually produces both **meningitis** and multiple parenchymal **abscesses** (granulomas) in patients with candidal sepsis; antecedent predisposing illnesses include severe burns, anatomic disruptions of urinary tract (such as hydronephrosis), peritonitis, or following prolonged broad spectrum antibiotic treatment

        3. **Rhinocerebral phycomycosis (zygomycosis; mucormycosis)** — **vasoinvasive** organisms spreading from **paranasal sinuses** into retro-orbital tissues and brain to cause fatal multifocal hemorrhagic brain infarction; usually associated with poorly controlled diabetic hyperglycemia and acidosis

        4. *Aspergillosis* — increasingly common infection found in **immuno-compromised individuals** (particularly in **AIDS**); due to embolization of **vasoinvasive** fungal hyphae from lung infection; fungal **emboli** obstruct brain vessels resulting in **infarcts** which then serve as culture media for organism proliferation

    D. Treatment consists of antifungal medication: amphotericin B, 5-fluorocytosine, or ketoconazole depending on specific organism sensitivity; however, in most cases unless underlying disease or immunologic compromise can be reversed, antifungal therapy is not successful

VII. **Neurosyphilis**

    A. Neurologic disease from infection by spirochete *Treponema pallidum* can occur in all patients with **inadequately treated syphilis**, producing various types of neurologic damage over many years; this complication is *completely avoidable by adequate early antibiotic treatment*

    B. Diagnosis based upon examination of cerebrospinal fluid

1. Nonspecific variable increases in cerebrospinal fluid cell count (lymphocytes and plasma cells), total protein level, and γ-globulin level; glucose levels are usually normal

2. Serologic tests

   a. Non-treponemal (reagin) antibody test — Venereal Disease Research Laboratory (**VDRL**) slide test negative in many cases of active neurosyphilis and may revert to negative in late disease; **biologic false positive** tests can be found with some inflammatory disorders (such as systemic lupus erythematosus or bacterial endocarditis)

   b. Specific treponemal antibody tests — fluorescent treponemal antibody absorption (**FTA-ABS**) test or *Treponema pallidum* immobilization (TPI) test positive in virtually all cases of neurosyphilis

3. With successful treatment, pleocytosis disappears first, followed by return of protein levels to normal; serologic studies revert to normal last

C. Syndromes of **neurosyphilis**

1. Asymptomatic neurosyphilis — no neurologic signs and symptoms; diagnosis based upon positive cerebrospinal fluid findings

2. **Meningovascular syphilis** — **strokes** due to **arteritis**; usually affects younger individuals; occurs 5-10 years after initial infection

3. **General paresis** (general paresis of the insane; dementia paralytica) — major cause of insanity prior to middle twentieth century, accounting for up to 10% of asylum admissions

   a. Progressive behavioral and **personality disturbances, dementia**, dysarthria, myoclonic jerks, seizures, spasticity, and **Argyll Robertson pupil**; occurs 10-15 years after initial infection; associated with chronic meningitis and hydrocephalus; progresses to death within 3-4 years if untreated; antibiotic treatment can halt progression, but usually does not improve neurologic disability

   b. **Argyll Robertson pupil** — pupils bilaterally are small and irregular, do not dilate to mydriatic drugs or constrict (react) to light, but constrict on accommodation ("accommodates, but doesn't react")

4. **Tabes dorsalis**

   a. Progressive **sensory loss** involving principally position and vibratory sense (**proprioception; dorsal column** sensation) with resultant **ataxia** (sensory ataxia); **positive Romberg test** (falling from standing position after eye closure)

# CENTRAL NERVOUS SYSTEM INFECTIONS

        b.    Lesser degree of loss of pain and temperature sense; associated with urinary bladder distention and overflow incontinence (from hypotonic insensitive bladder), trophic skin lesions, painless **perforating ulcers** (particularly in feet), and **joint destruction** from repeated injury to relatively anesthetic joints (neuropathic or **Charcot joints**; usually involving proximal joints such as knees and hips; in contrast to joint destruction with diabetic polyneuropathy which usually involves distal foot joints)

        c.    **Lightning pains** — brief sharp stabbing (lancinating) pains, most frequent in legs

        d.    Usually occurring 15-20 years after initial infection; treatment can prevent further sensory loss, but lightning pains, joint destruction, and gait disturbance from ataxia continue

    D.    Treatment involves high doses of penicillin and periodic monitoring of blood and cerebrospinal fluid serology

VIII.    **Lyme disease (*Borrelia burgdorferi*)**

    A.    Initial infection by spirochete follows ixodid tick bite (deer tick); initial flu-like illness (sometimes with **erythema migrans skin rash**)

    B.    Subsequent **chronic meningitis** and **radiculitis** can present as **facial palsy**, other cranial nerve palsies, or spinal root syndromes

    C.    **Chronic infection** produces **encephalitis** and **demyelination** that can mimic multiple sclerosis or neurodegenerative dementia

    D.    Diagnosis requires serologic studies of blood and cerebrospinal fluid; treatment involves administration of tetracyclines, penicillin, or cephalosporin antibiotics, in conjunction with periodic serologic monitoring

IX.    **Toxoplasmosis**

    A.    Minor illness in normal individuals caused by *Toxoplasma gondii*, but with immune compromise (such as in AIDS) can result in **meningoencephalitis** or **intracerebral mass lesion**

    B.    Since most adults have positive serologic titers, histologic identification of organism in brain biopsies is necessary to confirm diagnosis

    C.    **Congenital toxoplasmosis** — primary maternal infection during pregnancy results in fetal infection with consequent severe central nervous system damage; infant presents with **chorioretinitis, microcephaly** or hydrocephalus, **intracranial calcifications**, psychomotor retardation, and **hepatosplenomegaly** with jaundice

Chapter 13

X. **Cysticercosis**

  A. Multiple brain cysts produced by larval (embryo) form of **pork tapeworm** *Taenia solium*; larva migrate from intestine into circulation to be transported to muscle, eye, and brain; associated with intense inflammatory reaction and calcification

  B. Often asymptomatic; most common clinical symptom is seizures; radiologic imaging studies demonstrate multiple intracranial calcifications and ring-enhancing lesions; treatment with praziquantel (broad-spectrum antinematode drug) kills surviving organisms

XI. **Rickettsial infection**

  A. Abrupt onset of **fever, headache, myalgias, stiff neck**, confusion, **seizures**, lethargy and subsequent coma; associated with maculopapular rash developing two to four days after onset of fever; death results from pneumonia, renal failure, or circulatory collapse

  B. Pathology consists of generalized **necrotizing vasculitis** resulting in petechial hemorrhages and focal necrosis scattered throughout brain

  C. Causative agents include *Rickettsiae prowazekii* (**typhus**; transmitted by body lice); *Rickettsiae rickettsii* (**Rocky Mountain spotted fever**; transmitted by **ticks**), and *Rickettsiae tsutsugamushi* (scrub typhus; transmitted by chiggers)

  D. Diagnosis is based upon signs of neurologic disturbance associated with characteristic skin rash and history of insect bite; **Weil-Felix reaction** (antibody test) positive during second week of illness

  E. Treatment involves prompt administration of antibiotics, either intravenous chloramphenicol or oral tetracycline

XII. **Aseptic meningoencephalitis**

  A. Diagnosis based mainly on finding clinical signs and symptoms of **meningitis**, but with no identifiable causative bacterial, fungal, or parasitic organism; implies **viral etiology**, but meningeal inflammation without identifiable infecting organisms can be caused by introduction of foreign materials into subarachnoid space (such as intrathecal drugs) or by subarachnoid hemorrhage

  B. Cerebrospinal fluid shows increased cells (mostly **lymphocytes**), usually with **normal glucose level** and normal or slightly elevated protein level

  C. Presentation is usually of acute onset of **fever, headache, neck stiffness, irritability**, and **lethargy**; seizures and neurologic deficits (such as hemiparesis or cranial nerve palsies) indicates encephalitic component (brain parenchymal involvement)

D. Frequently occurs in **epidemics**; common viral etiologies include **mumps, enteroviruses** (coxsackie and ECHO), **lymphocytic choriomeningitis virus**, and **arthropod borne viruses**

E. Treatment is supportive once diagnosis is verified (finding no evidence of causative bacteria, fungus, or parasite); causative virus can be identified by serologic study comparing blood samples obtained during acute illness and three weeks later

XIII. **Herpes simplex encephalitis**

A. Subacute onset (over several days) of personality and **behavioral changes**, followed by rapid development of **fever, headache, seizures**, and **coma**

B. Radiologic imaging studies reveal **hemorrhagic necrosis of inferomedial temporal lobes** and **orbital frontal lobes**

C. Cerebrospinal fluid obtained by lumbar puncture shows increased pressure, increased number of white blood cells (mostly lymphocytes), many red blood cells (hemorrhagic necrosis), and elevated protein level; virus cannot be cultured from cerebrospinal fluid

D. Diagnosis confirmed by identification of organisms from **culture of brain biopsy**; **viral inclusions** can also be identified histologically and by electron microscopic study of brain biopsy

E. Herpes simplex encephalitis is **usually fatal** and survivors have profound dementia; **medical emergency** necessitating immediate treatment with **acyclovir** (Zovirax), since this drug has few side-effects; in many centers, if diagnosis of herpes simplex encephalitis is probable on clinical grounds, treatment is immediately begun with **acyclovir**, biopsy only being performed if patient does not become stable or improve within 48 hours; mortality and neurologic sequelae are reduced if treatment is instituted before onset of coma

XIV. **Poliomyelitis**

A. Acute febrile illness produced by poliomyelitis virus (**enterovirus** family); **initial gastrointestinal infection** is followed by **viremia** and subsequent infection of spinal cord **anterior horn motor neurons**; destruction of anterior horn cells results in **flaccid paralysis** (lower motor neuron weakness)

B. Mostly preventable by infant **immunization** programs using oral administration of mutant virus; occasional cases occur in unimmunized individuals or those without immunity due to immune system disorders

XV. **Rabies**

A. Subacute onset (over 3 to 4 days) of **personality change, hyperexcitability**, and **headache** followed by **dysphagia** and **pharyngeal muscle spasm** (leading to drooling or "frothing at the mouth" and inability to swallow water or "hydrophobia"), **facial muscle spasms, seizures**, and **coma**; nearly **universally fatal** after one to two weeks

B. Caused by virus transmitted to humans by **animal bites**; infected animals are diagnosed by finding characteristic **cytoplasmic inclusion (Negri body)** in cerebellar Purkinje cells

C. Incubation period varies from one to three months; virus is transferred to central nervous system by **axonal transport** in peripheral nerve motor and sensory axons terminating near area of bite; incubation periods are longer in leg bites and shorter in facial bites

D. Treatment is prophylactic with active and passive immunization

XVI. **Herpes zoster (shingles)**

A. Virus of **chickenpox** (varicella-zoster virus) resides asymptomatically in **sensory ganglia** (cranial nerve and dorsal root ganglia) following primary infection (chickenpox); **reactivation, produces inflammation and necrosis** in involved sensory ganglia and viral particles travel down axons to skin; viral multiplication in epidermal cells produces characteristic **vesicular rash** in distribution of involved sensory nerve (dermatomal distribution)

B. **Radicular pain and paresthesias** (particularly burning or itching) precede skin eruption by several days; recovery is often associated with decreased sensation in involved dermatome and chronic **pain (post-herpetic neuralgia)**

C. Viral reactivation presumably relates to reduced immunity; thus, elderly individuals have higher incidence, as do those with underlying **immune compromise** (such as with leukemia or lymphoma, immunosuppressive drugs, or cancer chemotherapy and radiotherapy)

D. Over 50% of cases involve thoracic dermatomes

E. Herpes zoster of trigeminal nerve (cranial nerve V) from reactivation of virus in **gasserian ganglion** results in **vesicular eruption on face**; involvement of first division of trigeminal nerve is associated with vesicular eruptions of conjunctiva and cornea producing **post-herpetic corneal scarring** (interfering with vision) and corneal anesthesia (with risk of perforation)

F. Treatment with acyclovir facilitates recovery of skin lesions, but has no effect on development of post-herpetic neuralgia

XVII. **AIDS (acquired immunodeficiency syndrome)**

A. **Multisystem disorder** produced by **human immunodeficiency virus (HIV)** infection; primary involvement of nervous system produces various neurologic syndromes

1. **Dementia** — progressive intellectual difficulties, **personality changes, tremor, myoclonus, and seizures**; often accompanied by **apathy, ataxia, disturbances of ocular motility, and hyperactive tendon reflexes**; late signs are **paraplegia** and **incontinence**

2. **Chronic basilar meningitis** — progressive **cranial neuropathies**, particularly involving trigeminal nerve, facial nerve, and vestibuloacoustic nerve

3. **Myelopathy** — moderate to severe **loss of posterior column sensation** (vibratory and position sense) and **spasticity** (due to corticospinal tract involvement) with bilateral **extensor plantar responses** (Babinski reflexes) despite reduction or loss of tendon reflexes; mimics vitamin $B_{12}$ deficiency

4. **Peripheral neuropathy** — progressive symmetrical **distal sensory loss**; related to combination of direct HIV infection and nutritional deficiency

5. **Inflammatory myopathy** — progressive **weakness**, elevated serum creatine phosphokinase (CPK) levels; histologic examination of muscle biopsy shows necrosis and inflammation

B. Complicated by **opportunistic infections** (such as cryptococcal meningitis, tuberculous meningitis, neurosyphilis, cerebral toxoplasmosis, aspergillus emboli with infarction, cytomegalovirus encephalitis, progressive multifocal leukoencephalopathy) and malignancy (primary central nervous system lymphoma or metastatic tumors)

XVIII. **Progressive multifocal leukoencephalopathy**

A. **Progressive neurologic disorder** occurring in setting of **immunologic compromise** associated with lymphoreticular malignancies (such as lymphoma or chronic lymphocytic leukemia), **AIDS**, or immunosuppressive drug treatment

B. **Rapid evolution of dementia, ataxia, visual field defects, spasticity and weakness** from (corticospinal tract damage), and **swallowing and speech difficulties**, followed by coma and **death usually within 6 months** of onset; no effective treatment is known

C. Pathologic change is **widespread destruction of white matter** which on microscopic examination shows **giant astrocytes with bizarre-shaped nuclei** and enlarged oligodendrocytes with inclusions

D. Causative agent is opportunistic **human polyoma virus, JC virus**

XIX. **Creutzfeldt-Jakob disease** (transmissible spongiform encephalopathy)

A. **Transmissible encephalopathy** related to kuru (disease of New Guinea highland natives), sheep scrapie, transmissible mink encephalopathy, and bovine spongiform encephalopathy ("mad cow disease")

B. **Rapidly progressive dementia** associated with **ataxia** and **myoclonus**

C. Electroencephalogram shows high voltage slow-sharp wave complexes on nearly flat background ("**burst suppression" pattern**)

D. Causative agent is **prion** which is composed of **protein only** (no nucleic acids); resistant to standard sterilization procedures (boiling, formalin, alcohol, or ultraviolet radiation), but inactivated by autoclaving or immersion in sodium hypochlorite (household bleach) or sodium hydroxide

E. Transmissibility possible from neural tissues; iatrogenic transmissions demonstrated after corneal transplantation, implantation in brain of infected recording electrodes, and injections of growth hormone extracted from cadaveric human pituitary; **dementia patients should not be used as organ donors**

XX. **Subacute sclerosing panencephalitis (SSPE)**

A. Slowly progressive degenerative disorder characterized by initial personality changes and **intellectual deterioration** with subsequent development of **myoclonic seizures, ataxia**, and **blindness**, progressing to death over several months or years

B. Markedly **elevated measles antibody titers** in cerebrospinal fluid

C. Develops several years after initial measles infection, apparently related to **reactivation** of latent form of **altered measles virus**; incidence has decreased in North America following widespread compulsory measles immunization, but is still prevalent in underdeveloped countries

XXI. **Tetanus**

A. Subacute onset (over days) of **generalized rigidity** beginning in jaws ("lockjaw"), neck, and back and progressing to **spasm** involving **all muscles**, including pharyngeal muscles and muscles involved in breathing

B. Symptoms result from **exotoxin** produced by gram-positive, spore-forming rod *Clostridium tetani*; initial symptoms appear several days to several weeks following **inoculation** of organism in **puncture or lacerating wound**; particularly common in older adults who have not been reimmunized for many years

C. **Tetanus neonatorum** — contamination of umbilical cord with organisms produces symptoms (within one week) of irritability, fever, feeding difficulty, muscular rigidity, cyanosis, and fatal apnea; can also occur with gastrointestinal colonization by spores from feedings of honey mixed with formula or water

D. Treatment involves antibiotics (penicillin or tetracycline), wound debridement to eliminate any toxin-producing organisms, antitoxin (human tetanus-immune globulin) administration, tracheostomy and ventilatory support, sedation, and use of muscle relaxants (high dose diazepam or *d*-tubocurarine)

E. Prevention is important by primary immunization with tetanus toxoid and booster immunizations at least every ten years

**Chapter 14**                                                                              **SPINAL COLUMN DISEASE**

I. Anatomic considerations

    A. Relatively small size of spinal cord (only about thumb size) packs many important structures in close proximity; it is also encased in relatively small space (vertebral canal)

    B. Spinal cord ends in conus medullaris at about L1-L2 vertebral level with cauda below that level; thus, lumbosacral vertebral disease tends to present as root symptoms, while cervical vertebral disease presents more often as spinal cord symptomatology

II. **Spinal anomalies**

    A. **Klippel-Feil anomaly** — congenital fusion of multiple vertebrae into single mass with reduced mobility and pain; associated with short neck and low hairline

    B. **Hemivertebrae** — can occur at any level; results in scoliosis; in thoracic level, usually associated with multiple congenital anomalies

    C. **Sacralization of L5** — fusion of L5 vertebrae to sacrum, often associated with pain

    D. **Spondylolysis and spondylolisthesis**

        1. **Spondylolysis** — **fracture** (occurring at or before birth) of vertebral **neural arch** separating superior and inferior articular processes (most often involves **L5**)

        2. **Spondylolisthesis** — in adolescent or young adult with spondylolysis, **forward slippage (subluxation)** of unstable vertebral body and superior portion of neural arch results in gradually increasing **back pain**, **limitation of motion**, **hamstrings spasm** (evident as **short stride gait**), and other signs of root compression; treatment involves surgical fusion

    E. **Odontoid deformities** — anomalies of odontoid result in **atlantoaxial joint instability** with pain and potential for spinal cord or medullary compression; frequently found in **Down syndrome**

    F. **Scoliosis — spinal curvature**

        1. Occurs in association with vertebral anomalies (such as **hemivertebrae**), **vertebral collapse**, **spinal tumors** (particularly those affecting nerve roots such as in **neurofibromatosis**), **spinocerebellar degenerations** (such as **Friedreich's ataxia**), **syringomyelia**, neuromuscular diseases (such as **spinal muscular atrophy** or **Duchenne muscular dystrophy**)

2. Occurs as **idiopathic** progressive disorder affecting mostly **young girls** (ages 6-14 years), involving **thoracic** vertebral levels

III. Visceral diseases — deep and burning pain radiating from abdomen or chest into back or neck sometimes mistaken for spinal disease; causes include **retroperitoneal masses or hemorrhage**, posterior wall myocardial infarction, posterior wall gastric ulcer, biliary or pancreatic disease, aortic aneurysm (particularly dissecting aneurysm), or pelvic tumors or inflammation

IV. **Inflammatory spine disease**

  A. Insidious onset of generalized **stiffness** and back and neck pain; **aggravated by maintaining one position** for long periods ("jelling effect"), such as after sleeping overnight in bed (**"morning stiffness"**); stiffness and pain are **relieved by mild activity**, but pain increases with more intense activities and may be greater in late afternoon or evening; **progressive limitation of back or neck motion**

  B. **Ankylosing spondylitis (Marie-Strümpell arthritis)**

   1. Neurologic signs minimal except in some cases with inflammation and fibrosis involving caudal canal and producing compression of nerve roots with consequent leg weakness, sensory loss, and loss of sphincter control

   2. Onset between ages 20 and 40 years, mostly in **men**

   3. Diagnosis confirmed by typical radiologic appearance of **destruction of sacroiliac joints** followed by **bridging of vertebral bodies**

   4. Treatment is symptomatic with anti-inflammatory drugs and physical therapy

  C. **Spinal rheumatoid arthritis (rheumatoid spondylitis)**

   1. **Destruction of upper cervical vertebrae** and associated ligaments can result in **atlantoaxial subluxation** with potentially fatal lower medullary and spinal cord compression

   2. Involvement of other joints (particularly hands) is helpful in establishing diagnosis

   3. Treatment is symptomatic with anti-inflammatory drugs, but immobilization with collars or surgical fusion may be required to prevent spinal cord compression

V. **Spinal infection**

  A. Progressive **unremitting back or neck pain**, often with focal tenderness to palpation or percussion over involved vertebral spine

B. **Vertebral osteomyelitis and spinal epidural abscess**

1. Results from **hematogenous** dissemination of distal infection, especially pelvic infections in elderly individuals; most common organisms are ***Staphylococcus aureus*, streptococci, and gram-negative bacteria**

2. Initial involvement of **vertebral body** results in **osteomyelitis**; subsequent spread into adjacent epidural fat produces **spinal epidural abscess**

3. Initial **localized tenderness** and **fever**, followed by headache, stiff neck, and rapidly progressive **paraplegia** (thoracolumbar localization) or **quadriplegia** (cervical localization)

4. Treatment involves immediate lesion localization with radiologic imaging studies, neurosurgical drainage and stabilization, and antibiotic pharmacotherapy

C. **Discitis (disc space infection)** — infection of intervertebral disc

1. Presents in children as slight fever, irritability, and refusal to sit, stand, or walk; focal tenderness evident on clinical examination

2. Diagnosis requires radiologic imaging studies; treatment involves bed rest and antibiotic pharmacotherapy

D. **Tuberculous osteitis (Pott's disease)** — *Mycobacterium tuberculosis* infection

1. **Pain and spinal deformity (gibbus or kyphosis)** usually involving lower thoracic area due to **granulomatous destruction** of vertebral bodies with subsequent collapse

2. Spinal cord compression can result in **paraplegia** and erosion through dura can produce **tuberculous meningitis**

3. Treatment involves immediate lesion localization with radiologic imaging studies, neurosurgical drainage and stabilization, and antituberculous pharmacotherapy

VI. **Meningeal irritation**

A. Acute or subacute **neck stiffness** and pain and pulling sensation in lower back (particularly with forward flexion of neck) associated with **hemorrhage** or **infection** in **subarachnoid space**

B. Patient tends to hold neck in hyperextension and guards against forward flexion (which increases pain in neck and between scapulae)

C. Associated with **fever** (particularly with infections) and **abnormal cerebrospinal fluid**

D. Treatment necessitates antibiotic therapy (infections) or identification of etiology (subarachnoid hemorrhage)

VII. **Intervertebral disc protrusion (herniation)**

A. **Lumbar disc**

1. **Low back pain with radiation down one leg**, often initially provoked by identifiable precipitating event (such as lifting heavy object); pain aggravated by movement and by coughing or straining; pain relieved by flexion of knees and thighs (lying on unaffected side with leg flexed)

| L4 Root Lesion | L5 Root Lesion | S1 Root Lesion |
|---|---|---|
| Difficulty with knee extension (quadriceps weakness) | Difficulty with foot extension (dorsiflexion) and foot drop | Difficulty standing up on tiptoes (calf muscle weakness) |
| Sensory loss from knee down medial aspect of lower leg to medial malleolus | Sensory loss over anterolateral lower leg down to dorsum of foot and big toe | Sensory loss over lateral aspect of heel and sole |
| Diminished or absent patellar tendon reflex (knee jerk) | Diminished or absent internal hamstring tendon reflex | Diminished or absent Achilles tendon reflex (ankle reflex) |

2. **Paravertebral muscle spasm** results in **pelvic tilt** (list) and **scoliosis** (convexity toward symptomatic side) with limitation of bending forward or backward

3. **Extremity weakness, numbness, and paresthesias** in localized areas of leg or foot depending on level of nerve **root compression**; occasionally (with midline rather than lateral disc herniation) associated with loss of sphincter control and sexual dysfunction; **positive straight leg raising test** (with patient lying prone, pain elicited by passive raising of straight leg)

4. Over 90% of herniated lumbar discs with nerve root compression occur at **L4-L5 interspace (affecting L5 nerve root)** or at **L5-S1 interspace (affecting S1 nerve root)**

5. Electromyography can provide objective evidence of **denervation** from **ventral (motor) root damage**; dorsal (sensory) root function can be evaluated by H-reflex and somatosensory evoked potential studies; radiologic imaging studies (myelogram, CT scans, MRI) reveal site of disk protrusion

6. Initial treatment involves **bed rest** on firm mattress along with **analgesic** and **muscle relaxant** medication; with resolution of pain, physical therapy including **back exercises** can be instituted; **chemonucleolysis** (enzymatic chymopapain dissolution of nucleus pulposus) can be used in some patients; surgical discectomy necessary for progressive motor deficit (pain alone usually does not respond to surgical intervention)

B. **Cervical disc**

1. **Neck pain with radiation into arm**; history of trauma or precipitating event only occasionally elicited; pain aggravated by upright posture and by coughing or straining; pain relieved by lying supine with arm abducted at shoulder

   | C6 Root Lesion | C7 Root Lesion |
   |---|---|
   | Weakness of elbow flexion (biceps) | Weakness of elbow extension (triceps) |
   | Sensory loss over thumb and lateral (radial) index finger | Sensory loss over index and middle fingers |
   | Diminished biceps and brachioradialis reflexes | Diminished triceps reflex |

2. **Limitation of range of neck motion** and **weakness, numbness**, and **paresthesias** in localized areas of arm or hand depending on level of nerve root compression; signs of **spinal cord compression** (spasticity in lower extremities) can be found with central (midline) disc protrusion

3. Over 80% of all herniated cervical discs with nerve root compression occur at **C5-C6 interspace (affecting C6 nerve root)** or **C6-C7 interspace (affecting C7 nerve root)**

4. Electromyography can provide objective evidence of denervation from ventral (motor) root damage; dorsal (sensory) root function can be evaluated by somatosensory evoked potential studies; radiologic imaging studies (myelogram, CT scans, MRI) reveal site of disk protrusion

5. Initial treatment involves cervical traction or cervical collar along with analgesic and muscle relaxant medication; early surgical discectomy is necessary for progressive motor deficit or any evidence of spinal cord compression

VIII. **Spinal stenosis**

A. **Lumbar stenosis**

1. **Back pain** with signs and symptoms indicating **compression of multiple bilateral lumbosacral nerve roots**; pain radiating into both lower extremities is provoked by exertion, by standing erect or by bending backward (typical complaint is pain upon reaching overhead that is relieved by bending forward); pain is relieved by sitting

2. Compression of lumbosacral nerve roots from **narrowing of lumbar spinal canal and nerve root foramina** by **degenerative vertebral changes** (disc space narrowing, nucleus pulposus protrusion and calcification, and osteophytic spur formation at vertebral body margins); changes superimposed on **congenitally narrow spinal canal**

3. Diagnosis is confirmed by radiologic imaging studies (myelography, CT scans, MRI)

4. Treatment includes wide surgical laminectomy with bilateral foraminotomy

B. **Cervical spondylosis**

1. Chronic nagging **centrally-located neck pain**, radiating to both arms with limitation of neck movement

2. Progressive weakness and wasting (**denervation atrophy** from nerve root compression) of upper extremity muscles and **spasticity** in lower extremities (**spondylitic myelopathy**)

    a. Must be distinguished from **amyotrophic lateral sclerosis** which has evidence of **widespread denervation** involving both upper and lower extremities along with spasticity

3. Degenerative process involving cervical vertebrae with **disc space narrowing**, nucleus pulposus protrusion and **calcification** (leading to **bar formation** bridging interspaces), and **osteophyte** (spur) formation at vertebral body margins; most frequent at C4-C5, C5-C6, and C6-C7 levels

4. Minor cervical trauma can precipitate sudden severe neurologic deficit (including paraplegia or quadriplegia) if spinal canal is significantly narrowed

5. Diagnosis is confirmed by radiologic imaging studies (myelography, CT scans, MRI)

6. Treatment includes cervical traction or cervical collar along with analgesic and muscle relaxant medication; decompression of spinal cord and nerve roots is indicated for persistent radicular pain with neurologic deficit or for significant spinal cord compression

IX. **Spinal tumor**

A. Steady relentless progression of pain and other symptoms which depend on level and site of tumor

B. **Extramedullary tumors**

1. Growth **outside spinal cord parenchyma** produces symptoms related to nerve root compression and bone destruction before spinal cord symptoms

2. Usually results from **local spread of vertebral body metastases** to epidural space; most common primary cancers metastasizing to thoracic or lumbar vertebrae are **lung, breast**, and **prostate**

3. Initial symptoms include **unremitting** severe localized back or radicular **pain**, weakness or numbness in legs, and urinary retention; rapidly progresses to complete **paraplegia** (from vascular compression with consequent spinal cord infarction)

4. Treatment involves administration of **high-dose corticosteroids** to reduce edema and surgical decompression or radiotherapy to reduce tumor bulk and relieve pressure on spinal cord

C. **Intramedullary tumors** — growth **within spinal cord parenchyma** producing early spinal cord symptoms from local destruction and disruption of long tracts

D. Radiologic imaging studies confirm location of neoplasm which may require surgical biopsy for diagnosis unless primary cancer is already known

X. **Spinal cord infarction (myelomalacia; anterior spinal artery syndrome)**

A. **Acute onset** (often over minutes, but usually within several hours) of **paralysis** (paraplegia or rarely quadriplegia) and **dissociated sensory loss (loss of pain and temperature** sense below lesion level due to spinothalamic tract damage, but **preservation of proprioception** due to sparing of posterior columns)

B. Most often associated with severe **aortic atherosclerosis** (particularly aortic aneurysms) compromising origins of radicular arteries which contribute to anterior spinal artery; largest of radicular vessels (**great anterior medullary artery of Adamkiewicz**) originates at about **T11 level** and provides most of blood supply for lower two-thirds of spinal cord

XI. **Syringomyelia**

A. Slowly progressive **upper extremity muscle weakness and wasting** along with **dissociated sensory loss** (loss of pain and temperature sense, but preservation of proprioception); symptoms can be present for many years and only accidently discovered (such as following painless injury or burn)

B. Sensory loss usually has **cape-like distribution** over arms, upper trunk and back, neck, and back of head

C. Slight degree of spasticity and ataxia can be detected in lower extremities; occasional individuals have unilateral or bilateral Horner's syndrome (ptosis, miosis, and anhydrosis)

D. Pathologic changes — **central glial-lined cavitation of cervical spinal cord**; cavity may extend upward into medulla and pons (**syringobulbia**) or downward into thoracic and lumbar spinal cord

E. Secondary causes of cervical spinal cord cavitation include spinal cord astrocytoma, postradiation myelopathy, spinal cord infarction, and intraparenchymal spinal cord hemorrhage (hematomyelia)

F. Diagnosis requires radiologic imaging studies

# SPINAL COLUMN DISEASE

G. Treatment involves surgically opening cavity to subarachnoid space

XII. **Psychogenic pain**

A. Characterized by **inconsistent** description of pain along with normal neurologic examination

B. Often related to injuries for which litigation is pending

C. Treatment requires appropriate diagnosis and management of underlying psychiatric condition or emotional factor

# Chapter 15  SLEEP DISORDERS

I. **Sleep disorders** are extremely common, affecting possibly up to 10% of North American population

II. Normal **sleep physiology** — two basic categories of sleep defined by electroencephalography (EEG) and other physiologic parameters:

   A. **Rapid eye movement (REM) sleep**

   1. EEG characteristics of drowsiness or waking, but with **bursts of eye movement**

   2. **Inhibition of muscle tone** (no recordable electrical activity in muscles), except in extraocular and diaphragm muscles

   3. **Irregular breathing**, bursts of **sympathetic autonomic activity** (including tachycardia, hypertension, and pupillary dilation), **penile tumescence** (erection), and **increase in body temperature** and metabolic rate

   4. Associated with **dreaming**

   B. **Non-REM (NREM) sleep**

   1. Divided into **four stages — drowsiness (stage I), spindle sleep (stage II)**, and **slow wave or deep sleep (stages III and IV)**

   2. **Decreased body temperature** and metabolic rate, **bradycardia**, slow breathing rate, **increased vagal tone**

### CHARACTERISTICS OF SLEEP STAGES

| CATEGORY | STAGE | APPEARANCE | PHYSIOLOGY | | ABNORMALITY |
|---|---|---|---|---|---|
| rapid eye movement (REM) | REM | very still with loss of tone; visible eye movements | REM bursts; loss of EMG activity; irregular breathing; autonomic sympathetic bursts and penile tumescence; increased body temperature | | narcolepsy; nightmares |
| non-REM | I | drowsiness | myoclonic jerks | lower body temperature; slow pulse and breathing; EMG activity; increased vagal activity | sleep myoclonus; seizures |
| | II | light sleep | sleep spindles | | |
| | III | deep sleep | slow wave sleep | | sleepwalking; night terrors; enuresis |
| | IV | deep sleep | slow wave sleep | | |

(a) **Normal muscle tone** and **frequent movements** (prominent myoclonic jerks at onset of stage I)

3. **Secretion of growth hormone**, particularly during stages III and IV sleep

C. **Sleep stage effects on underlying medical disorders**

1. **REM sleep** bursts of sympathetic autonomic nervous system activity can exacerbate **coronary angina**

2. **Increase in epileptiform EEG activity** in epileptics during **stage I** sleep (drowsiness) predisposes to clinical seizures

3. **REM sleep** increase in **gastric acid secretion** can exacerbate gastritis or ulcer pain

D. **Sleep Cycle**

1. Adult sleep cycle begins with stage I, progresses through stage II, then stages III and IV, and back through stages II and I, until about 70 to 100 minutes after sleep onset, first REM stage begins; subsequently, this cycle repeats about every 90 minutes (four to six cycles per night), but during later cycles, REM periods lengthen and there is much less (or no) stages III and IV sleep

2. Sleep cycle changes with age

    a. Length of total sleep varies: from about 16 hours in full-term newborn infants, to 12 hours at age 12 months, 8 hours at puberty, and 6 hours at age 50 years

    b. REM sleep constitutes 20% of total sleep from about age 3 years (adult pattern); in full-term newborn infants, REM sleep constitutes over 50% of total sleep and often precedes NREM sleep

    c. Stages III and IV sleep decrease with age, such that by about age 60 years there is no stage IV and little stage III sleep

    d. Frequent nighttime awakenings are common in older individuals

III. **Biologic rhythms**

A. Terminology

1. **Circadian rhythms** — rhythms with cycles **approximating one day (24 hours)**

2. **Ultradian rhythms** — rhythms with cycles **less than 24 hours**

3. **Infradian rhythms** — rhythms with cycles **greater than 24 hours**

B. **Body temperature** — falls in late afternoon, lowest near end of sleep period, then rises just before morning awakening

C. **Hormonal secretion**

1. **Growth hormone** — secretion during sleep, mainly during first **NREM** period of **stages III and IV sleep**; greatest secretion in children, decreasing beyond adolescence, and nearly absent in elderly individuals

2. **Prolactin** — secretion mostly just before morning awakening; continuous secretion (loss of circadian fluctuation) follows treatment with dopamine-blocking neuroleptic drugs (such as phenothiazines) and during pregnancy and lactation; treatment with dopamine-agonist drugs (such a bromocriptine) suppresses secretion

3. **Adrenocorticotropic hormone (ACTH)** and **cortisol**

   a. Bursts of ACTH secretion are greatest in early morning with cortisol levels rising after ACTH bursts (plasma cortisol levels highest between 6:00 am and 8:00 am; no release of cortisol between 10:00 pm and 2:00 am, with lowest plasma levels at about 2:00 am)

   b. **Dexamethasone suppression test**

   (1) In normal individuals, single oral dexamethasone dose at night will suppress morning ACTH burst and cortisol production

   (2) Loss of normal circadian cortisol rhythm is found in **depressive disorder** (endogenous depression); cortisol production is unaffected by dexamethasone administration (**non-suppression** of cortisol release)

IV. **Parasomnias** — conditions occurring exclusively during sleep or exacerbated by sleep

A. **Bed-wetting (enuresis)** — in absence of urologic disorder, bed-wetting beyond age 5 years is abnormal; can occur during all sleep stages, but more common during **stages III and IV sleep** (deep sleep)

1. **Primary enuresis** — wetting nearly every night (daily or several times per week) from infancy, but normal daytime bladder control

   a. Most often related to **maturational lag**, with **positive family history** of enuresis

# SLEEP DISORDERS

      b.    Spontaneous resolution: 10% of 6 year old children, but only 1% of 18 year old adolescents have enuresis

  2.  **Secondary enuresis — relapse** to bed-wetting after dry period of several months to years

      a.    Often due to **psychological factors** such as childhood depressive disorder or medical conditions such as **urinary tract infection, diabetes mellitus, diabetes insipidus**, or **epilepsy**

  3.  Sometimes treatment with low doses of imipramine (Tofranil), a tricyclic antidepressant, reduces frequency of bed-wetting; behavioral techniques (such as bladder control exercises, rewards for dry beds, and wetting alarms) also can be beneficial

B.  **Sleepwalking (somnambulism)**

  1.  Complex behaviors initiated during **stages III and IV sleep** (deep sleep), lasting several minutes; individual often returns to bed without awakening, but if awakened will be confused and have no recall of episode

  2.  Behavior ranges from sitting-up in bed to walking, running, or climbing, and can include talking

  3.  Most common between ages 4 and 8 years; usually disappears spontaneously after adolescence; occurs in up to 15% of children

  4.  Individual must be protected from injury by locking doors and windows and avoiding other dangerous situations

C.  **Night terrors (pavor nocturnus)**

  1.  Sudden screams or cries, accompanied by extreme autonomic discharge (tachycardia, tachypnea, mydriasis, and sweating) suggesting intense fear; occurs **during stages III and IV sleep (deep sleep)**

  2.  Difficult to awaken, but if awakened individual is usually confused and amnestic for episode (sometimes single frightening image or sense of doom can be recalled)

  3.  Most common in children and usually disappears by adolescence; frequently associated with sleepwalking

D.  **Nightmares**

  1.  Long complicated frightening dreams occurring in **REM sleep** that awaken individual from sleep

2. Nearly 50% of children have nightmares at sometime before age 6 years

E. **Nocturnal (sleep) myoclonus**

1. Abrupt **repetitive jerking of feet and legs** occurring during **drowsiness (stage I sleep)** and **spindle sleep (stage II sleep)**; often associated with brief awakening and can be violent enough to disturb bed partner

2. Associated with excessive daytime sleepiness due to fragmented restless sleep

3. Treatment with clonazepam has been reported to reduce number of arousals

F. **Restless legs syndrome**

1. **Indescribable, unpleasant sensations in legs** resulting in **irresistible urge to move**, with movement seeming to relieve discomfort

2. Unpleasant sensations in legs are often worse in evening and on attempting to go to sleep

3. Frequently associated with nocturnal myoclonus

4. Familial occurrence in up to one-third of cases; etiology not known, but sometimes associated with anemia, sensory neuropathy, or uremia

5. No specific treatment has proven effective; sedative-hypnotics can provide some relief

V. **Insomnia** — difficulty falling asleep or staying asleep

A. **Transient insomnia (situational insomnia)**

1. Change in sleep pattern associated with precipitating cause (such as death of loved one, divorce, change of job, test, or sleeping in unfamiliar surroundings)

2. More common in individuals who are insecure and prone to emotional arousal (anxiety)

B. **Chronic (persistent) insomnia**

1. Persistent insomnia, often seeming to evolve from situational insomnia

2. Can be differentiated from normal individuals who need little sleep by evidence of **decreased daytime energy** and **attention** with chronic insomnia

3. Often associated with psychiatric disorders and secondary maladaptive behaviors such as poor sleep hygiene (behaviors that produce arousal such as drinking caffeine-containing beverages before bedtime)

C. Insomnia associated with psychiatric disorders — anxiety, manic-depressive disorder, schizophrenia, and other psychiatric illnesses are often associated with insomnia

D. Insomnia associated with drugs and alcohol (**drug-rebound insomnia**)

1. Chronically-administered **sedative-hypnotics** (such as benzodiazepines, barbiturates, glutethimide, or chloral hydrate) lose effectiveness (tolerance) after several weeks, necessitating increased dosage and resulting in poor sleep, daytime drowsiness, poor coordination, ataxia, slurred speech, and restlessness; abrupt withdrawal leads to marked inability to fall asleep

2. **Stimulant drugs** (caffeine-containing beverages, ephedrine, amphetamines, or weight-reducing drugs) can increase arousal and lead to insomnia

3. **Chronic alcohol abuse** interferes with normal sleep cycle resulting in insomnia and daytime drowsiness

VI. **Narcolepsy**

A. Disorder of **excessive sleepiness** associated with **cataplexy, sleep paralysis,** and **hypnagogic hallucinations**

1. **Cataplexy — sudden loss of muscle tone** (can fall without warning); often **precipitated by laughter** or any strong emotion, while fully awake; recovery is immediate

2. **Sleep paralysis — inability to move even though awake**; occurs during transition between sleep and waking and lasts up to several minutes before ability to move returns; often very frightening and associated with feeling of being unable to breath

3. **Hypnagogic hallucinations** — vivid (often frightening) **dream-like** visual or auditory perceptions occurring at sleep onset

B. Irresistible "sleep attacks" occur repeatedly and last less than 15 minutes, after which individual awakens feeling refreshed

C. Physiologic tests

1. **Polysomnography** (physiologic recording of sleep including EEG) shows **reduced sleep latency** (shortened interval before onset of sleep) and **sleep-onset REM** period (REM sleep

Chapter 15

occurring soon after falling asleep and often associated with report of sleep paralysis or hypnagogic hallucinations)

2. Multiple sleep latency test (MSLT) shows **frequent sleep attacks** with sleep-onset REM periods

D. Usual **onset in adolescence**; **positive family history** for similarly affected individuals

E. Associated with **HLA-DR2 phenotype** (most narcolepsy patients from all ethnic groups have this phenotype; however, relationship is unclear, since many individuals without narcolepsy also have this phenotype)

F. Treatment involves **stimulant medications** such as methylphenidate (Ritalin) which reduce sleep attacks; cataplexy responds best to tricyclic antidepressants such as imipramine (Tofranil), protriptyline (Vivactil), or clomipramine (Anafranil)

VII. **Sleep Apnea**

A. **Central (neurogenic) sleep apnea**

1. Apneic periods lasting up to 30 seconds, followed by resumption of breathing (often with hyperventilation) and then another apneic period; occurs mainly during transition from waking to sleep and when in supine position

2. Hypoxemia (blood oxygen desaturation) during sleep

3. Associated symptoms and complications include cardiac arrhythmias, right heart failure, hypertension, night sweats, morning headache, and excessive daytime sleepiness.

4. Produced by any central nervous system disorder interfering with medullary respiratory center function

B. **Obstructive sleep apnea**

1. **Sleep-related upper airway obstruction** from combination of:

a. Upper airway narrowing — due to **bulky soft tissues** (such as **enlarged tonsils and adenoids**, **hypothyroidism**, or **acromegaly**) or **craniofacial anomalies** (such as **micrognathia** or **macroglossia**)

b. Abnormal neural control of nasopharyngeal musculature that accentuates normal relaxation of muscles around upper airway during sleep

2. Characterized by periods of **loud snoring and snorting** (often disturbing bed partner) alternating with apneic periods; during apneic periods (usually lasting about 10-40 seconds) there is no airflow despite repeated inspiratory efforts and diaphragm movement, after which markedly increased breathing efforts result in restoration of airflow accompanied by loud snoring or snorting sounds and brief arousal

3. Apneic episodes occur during **REM sleep** periods

4. Full arousal during sleep may be difficult, but if awakened individual often is confused; common presenting complaints include being unrefreshed upon morning awakening, excessive daytime sleepiness, and morning headaches

5. Hypoxemia (blood oxygen desaturation) during sleep

6. Cardiovascular complications include systemic and pulmonary arterial hypertension, cor pulmonale, **polycythemia**, and **right heart failure**

7. Often associated with obesity and most common in **obese, middle-aged men**

    a. **Pickwickian syndrome** — term previously used to describe obese males with obstructive sleep apnea accompanied by daytime sleepiness and cardiovascular complications including polycythemia; term derived from description of fat, sleepy, red-faced boy in *The Pickwick Papers* by Charles Dickens

8. Treatment includes weight loss and surgical treatment of obstruction (including tracheostomy); sedative-hypnotic drugs should be avoided

VIII. **Circadian rhythm sleep disorders**

A. **Jet lag** (time zone change syndrome) — insomnia and drowsiness due to delay in adjustment of endogenous circadian rhythms to new environmental time cues; symptoms spontaneously abate after two to four days (longer after eastward flights)

B. **Shift work sleep disorder** — insomnia and drowsiness due to changes in work schedule necessitating sleep during usual waking hours; symptoms spontaneously improve after maintenance of same schedule for about one week; however, optimum alertness during waking and restful sleep generally only occur when work schedule allows sleep at night and waking during daytime

**Chapter 16**                                    **MULTIPLE SCLEROSIS AND DEMYELINATION**

---

I. **Multiple sclerosis (MS)** — most common **demyelinating disease** of central nervous system

    A. Characterized by **relapsing and remitting attacks** involving **multiple locations** within nervous system ("lesions disseminated in space and time")

    B. Signs and symptoms relate to whatever part of brain or spinal cord is damaged during particular episode

        1. Common initial symptoms include **visual impairment** (including blindness), **limb weakness**, **diplopia**, vertigo, ataxia, dysesthesias, emotional lability, impotence, and bladder dysfunction; symptoms often are transient or fluctuate in severity, and seem bizarre (and unable to be confirmed objectively by physician examiner), particularly early in disease process

        2. **Lhermitte's symptom** — sudden, electric shock-like, tingling paresthesias in neck radiating into arms and/or legs produced by neck flexion; indicative of lesion of cervical portion of spinal cord dorsal (posterior) columns

        3. Common later symptoms include generalized weakness and spasticity, pain, hearing loss, dysarthria and swallowing difficulties, and dementia

        4. Initial visual abnormalities (often unilateral) occur in 25% of patients

            a. **Optic neuritis** — **swelling of optic nerve (papillitis)** visible on funduscopic examination; associated with marked **loss of visual acuity**; must be differentiated from papilledema (due to increased intracranial pressure) in which there is minimal or no change in visual acuity

            b. **Retrobulbar neuritis** — **loss of visual acuity** with no identifiable funduscopic abnormality

            c. Mnemonic for difference between optic neuritis and retrobulbar neuritis: "in optic neuritis, physician sees something [papillitis] and patient sees nothing [blindness]; in retrobulbar neuritis, physician sees nothing [optic disc normal] and patient sees nothing [blindness]"

            d. Objective evidence of visual loss includes abnormal visual evoked potentials and **afferent pupillary defect** (pupil does not constrict with light shown into blind eye, but does constrict to light shown into contralateral normal eye)

e. Most patients show improvement of visual function (some completely recover), but recurrent or severe attacks result in optic disc pallor and atrophy with permanent visual impairment

5. **Diplopia and nystagmus** are commonly due to **internuclear ophthalmoplegia**

   a. Internuclear ophthalmoplegia superficially resembles medial rectus muscle paralysis except that **ability to converge (look at near objects) is spared**; with attempts to look laterally, *ab*ducting eye moves laterally and develops nystagmus (rhythmic involuntary to-and-fro oscillations of eye), while *ad*ducting eye does not move past midline

   b. Results from lesion of **medial longitudinal fasciculus** in central part of pons on same side as inability to adduct eye

6. Other signs (in order of frequency) include weakness, spasticity and hyperreflexia, Babinski reflex (extensor plantar response), intention tremor, ataxia, urinary incontinence, dysarthria or scanning speech (variable intonation and involuntary interruption often with explosive quality), and altered sensibility

7. Uhthoff phenomenon — worsening of neurologic signs or symptoms due to elevation of body temperature (such as with fever, saunas, or hot bath); has been used in past as diagnostic test ("hot bath test") for multiple sclerosis, since neurologic signs may become obvious by immersing patient in hot bath

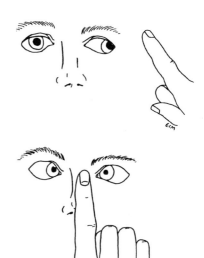

Internuclear ophthalmoplegia: adducting eye cannot go past midline, while abducting eye shows nystagmus; differentiated from medial rectus palsy by normal convergence.

C. No pathognomonic tests exist; diagnosis requires correlation of history, examination findings, and laboratory studies to demonstrate pattern of relapsing and remitting neurologic signs and symptoms involving multiple central nervous system areas associated with laboratory findings consistent with demyelination

1. **Magnetic resonance imaging** — multiple areas of **white matter demyelination**, particularly in **periventricular** regions

2. Cortical evoked potentials — abnormalities in visual, auditory, and somatosensory evoked potentials indicate lesions in myelinated fiber pathways in most patients; visual evoked potentials are abnormal in more than 75% of patients, brain stem auditory evoked responses in 50%, and somatosensory evoked potentials in 70%, even without symptoms

3. **Cerebrospinal fluid (CSF)** examination — lumbar puncture samples show normal or slightly increased cell count, normal or slightly increased protein (but not more than 100 mg/dL), and slightly increased total γ-globulin level (greater than 12% of total CSF protein); relatively specific findings include:

   a. IgG index — ratio of CSF and serum levels of IgG and albumin; elevation indicates increased synthesis of immunoglobulins inside blood-brain barrier

   b. **Oligoclonal bands** — several distinct bands of immunoglobulin in high-resolution gel electrophoresis of CSF in nearly 90% of patients with multiple sclerosis (occasionally also seen in patients with encephalitis or meningitis)

   c. Myelin basic protein — elevated levels indicate active demyelination within central nervous system

D. Course is unpredictable and highly variable, but about 20% have "benign" form (few exacerbations with nearly complete remissions and minimal disability), 30% "exacerbating-remitting" form (many exacerbations but long intervening periods of stability and some disability), and 50% "chronic relapsing/progressive" form (many exacerbations with few short remissions and increasing disability)

E. Neuropathologic changes — relatively sharply circumscribed **plaques** in which there is complete or nearly complete **destruction of myelin** (with relative **preservation of axons**), proliferation of reactive astrocytes, and accumulation of macrophages particularly in perivascular spaces

F. Etiology is unknown but seems to be related to interaction between genetic propensity and environmental factors affecting immunologic functioning

   1. **Experimental allergic encephalomyelitis (EAE)** — often considered immunologic **model** of multiple sclerosis; animals injected with central nervous system myelin or various myelin components (such as myelin basic protein or proteolipid protein) dissolved in Freund's adjuvant (an immune system stimulant composed of paraffin oil, killed mycobacterium, and emulsifier) develop either monophasic inflammatory demyelination (older animals) or relapsing-remitting demyelinating illness (younger animals)

   2. Epidemiology

      a. **Unequal geographic distribution, with increased frequency closer to each pole and decreased frequency nearer equator**; incidence is 1 per 100,000 in equatorial areas, 14 per 100,000 in southern United States, and nearly 80 per 100,000 in northern United States and Canada

      b. Immigrant population studies indicate risk can be altered by migrating from high incidence area to low incidence area: migration before age 15 years reduces risk to that of new area, but migration after age 15 years retains higher risk of original homeland

3. Certain histocompatibility types are more common in multiple sclerosis victims than in general population (particularly HLA-DR2)

4. Increased levels of immunoglobulin, free κ-light chains, and increased numbers of T-lymphocytes found in cerebrospinal fluid

G. Treatment — **no curative therapy** is currently available, but symptomatic treatment and management of disabling complications are possible

1. **Immunosuppressive agents** such as corticosteroids (prednisone), adrenocorticotropic hormone (ACTH), cyclophosphamide, cyclosporine, azathioprine, or interferon have all been used to shorten duration of acute exacerbations and extend periods of remission

2. **Bladder hygiene** (including self-catheterization for residual urine) is necessary to prevent or treat urinary tract infections caused by bladder dysfunction

3. **Antispasticity agents** such as baclofen (Lioresal), dantrolene (Dantrium), or diazepam (Valium) can reduce spasticity, but can also substantially increase weakness

4. Pseudobulbar affect (reflex spontaneous laughter or crying due to damage to corticobulbar pathways) and depressive symptoms respond to low doses of tricyclic antidepressants such as amitriptyline (Elavil)

5. Fatigue can improve with amantadine (Symmetrel)

6. **Physical and occupational therapy** can help to maintain as much function as possible

II. **Acute disseminated encephalomyelitis**

A. **Monophasic, self-limited demyelinating disease**

1. **Postvaccinal** encephalomyelitis — develops one to two weeks following vaccination against smallpox or rabies (particularly with brain-derived vaccine)

2. **Postinfectious** encephalomyelitis — develops two to six days following onset of viral exanthem (such as measles, chickenpox, smallpox, or rubella) or rarely preceding exanthem

B. **Abrupt onset** of **headache**, **fever**, **confusion**, and **stiff neck**; severe cases can progress to convulsions, cerebellar ataxia, quadriplegia, cranial nerve palsies, or coma; **acute cerebellitis (acute cerebellar ataxia)** form most common following chickenpox (varicella)

C. Laboratory studies are not specific, but suggest inflammatory disease involving white matter

    1. **Magnetic resonance imaging — widespread white matter demyelination**

    2. Cerebrospinal fluid studies — slightly increased number of lymphocytes (up to 250 cells/$\mu$L), slightly increased protein (up to 150 mg/dL), and increased levels of myelin basic protein

D. Mortality is high (up to 50% in postvaccinal form); recovery can be complete or leave significant residual neurologic abnormalities

E. Treatment with immunosuppressive agents (corticosteroids or ACTH) can reduce severity of illness; anticonvulsants are sometimes necessary for control of seizures; mannitol is often required to reduce cerebral edema

F. Pathologic findings — **perivenous demyelination** (with relative sparing of axons) associated with large numbers of lymphocytes and lipid-laden macrophages

G. Adult form of experimental allergic encephalomyelitis is thought to be model of acute disseminated encephalomyelitis

H. **Acute necrotizing hemorrhagic encephalomyelitis — fulminant form** of disease with widespread white matter necrosis, marked CSF leukocytosis (up to 3000 cells/$\mu$L), and increased intracranial pressure; usually fatal in two to four days; must be differentiated from herpes simplex encephalitis which tends to be localized to temporal lobes

III. **Acute transverse myelitis** — syndrome with multiple causes; characterized by **acute** (over hours to several days) or **subacute** (one to two weeks) **spinal cord damage**, usually involving **thoracic level**

  A. Initial presentation is with localized back pain (usually thoracic radicular pain) followed by abrupt onset of weakness and loss of sensation in both legs, loss of bladder and anal sphincter function, and subsequent development of **thoracic sensory level and paraplegia**

  B. Causes include **postinfectious** or **postvaccinal** syndrome, presumed **toxic injury** (especially related to intravenous drug abuse), direct viral injury, or vasculitis (with occlusion of spinal cord vasculature)

IV. **Central pontine myelinolysis**

  A. Rapidly progressive **corticospinal and corticobulbar syndrome** with flaccid quadriplegia and pseudobulbar palsy (upper motor neuron facial and oropharyngeal weakness); often fatal within days or weeks

## MULTIPLE SCLEROSIS AND DEMYELINATION

B. Initially described in chronically debilitated alcoholics; now recognized to result from **overly rapid (iatrogenic) correction of hyponatremia and/or serum hyperosmolality** (from any underlying cause including malnutrition, chronic hepatic or renal disease, leukemia, alcoholism, or infections)

C. MRI reveals demyelination centered in pontine median raphe involving all or part of basis pontis

D. Neuropathologic findings — myelin destruction (myelinolysis) in ventral pontine tegmentum and basis pontis with preservation of axons and neurons of pontine nuclei and without any inflammatory cells; similar myelin destruction sometimes extends into internal capsule and cerebral white matter

E. **Prevention** is necessary by **judicious use of fluids** to slowly correct hyponatremia or serum hyperosmolality in hospitalized patients

Chapter 17  DEVELOPMENTAL DISORDERS

I. **Developmental milestones and signs**

   A. **Moro reflex** — symmetrically present in term newborns, and disappearing by age 3 months; **asymmetry** indicates central nervous system disease such as hemiparesis, brachial plexus injury, or spinal cord lesion

   B. **Tonic neck reflex** — appears by age 1 month and disappears by age 5 months; **obligate reflex** (infant unable to move out of posture) indicates brain disease and precludes rolling over; must disappear prior to ability to sit

   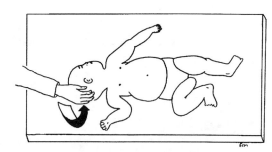

   Tonic neck response elicited by turning head: arm (and leg) extend on side toward which head is turned while other arm (and leg) flex ("fencing posture").

   C. **Traction response** — by age 6 months head should no longer lag when infant is pulled to sitting position; lack of head control (**head lag**) is evident in hypotonic (floppy) infants

   D. **Parachute reflex** — appears at 6 months (and persists throughout life); **asymmetry** suggests hemiparesis, spinal cord lesion, or brachial plexus lesion.

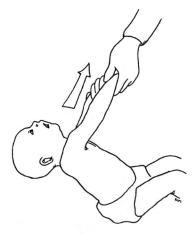

Head lag should disappear by age 6 months.

| SELECTED CAUSES OF FLOPPINESS |
|---|
| **Without Weakness** |
|    Perinatal hypoxic-ischemic encephalopathy |
|    Chromosomal disorders |
|    Prader-Willi syndrome |
|    Hypothyroidism |
|    Amino acidopathies |
|    Lipid storage disorders |
| **With Weakness** |
|    Spinal cord transection |
|    Spinal dysraphism |
|    Werdnig-Hoffman disease |
|    Congenital myotonic dystrophy |
|    Neonatal myasthenia gravis |
|    Infantile botulism |
|    Pompe's disease (acid maltase deficiency) |

# DEVELOPMENTAL DISORDERS

E. Horizontal suspension — hypotonic (floppy) infants are unable to arch back or hold head above horizontal by age 5 months

F. Righting reflex — sitting propped should be achieved by age 4 months and sitting alone by age 6 months; symmetrical righting reflexes should be present in sitting position; asymmetry or absence suggests cerebral or spinal cord lesions

G. Walking — **independent walking achieved by age 15 months**; delay beyond age 18 months indicates brain, spinal cord, or neuromuscular abnormality

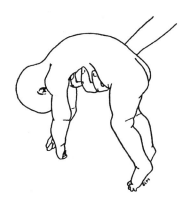

Inverted U-shape of floppy infant unable to arch back or hold head above horizontal by age 5 months.

H. Language — milestones include smiling responsively by age 8 weeks, playing "peek-a-boo" and babbling single words ("dada," "mama," "bye") by age 12 months, using single words by age 15 months, and by age 24 months using 2-word phases, pointing to objects, and obeying simple commands; delayed language suggests central nervous system disease, hearing or visual deficit, or autism

II. **Cerebral palsy (static congenital encephalopathy)**

A. Group of disorders with **prominent motor deficit** (delay in normal motor milestones) secondary to brain lesion

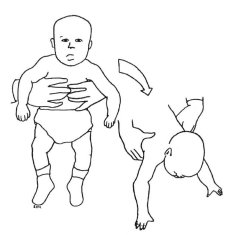

Parachute reflex is elicited by plunging suspended infant downward: arms should thrust forward symmetrically as though to break the fall.

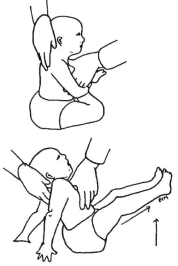

At age 6 months, infant pushed backward should kick out legs symmetrically.

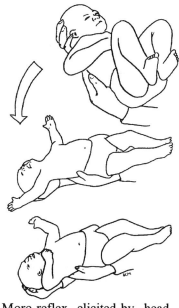

Moro reflex elicited by head extension has two phases: extension and abduction of arms and leg extension, followed by slower adduction of arms.

167

acquired prenatally or perinatally, usually results from hypoxic-ischemic insult, infection, hemorrhage, brain maldevelopment; intellectual ability may be spared

    B. Specific patterns include:

        1. **Spastic diplegia — spasticity in both legs** with relative sparing of arms; usually related to bilateral hypoxic-ischemic lesions in parasagittal "border zone" (area at border between vascular supplies of major cerebral arteries)

        2. **Spastic hemiplegia — unilateral** cerebral hemispheric damage (often related to trauma, vascular event, or maldevelopment); **spasticity in arm and leg on same side** with undergrowth of limbs and abnormal posturing

        3. **Choreoathetosis** — initial hypotonia evolving into choreoathetosis and dystonia; now rare, but previously common secondary to **kernicterus**

    C. Treatment involves ruling out degenerative diseases and neoplasms, maximizing mobility through physical therapy, occupational therapy, and orthopedic surgery, and arranging for intellectual assessment for proper school placement

III. **Congenital intrauterine infections**

    A. **TORCHES** (**TO**xoplasmosis, **R**ubella, **C**ytomegalovirus, **HE**rpes simplex, **S**yphilis) and **human immunodeficiency virus** (HIV)

    B. Diagnosis established by serologic study of **cord blood** showing elevated disease-specific **IgM level**

    C. Typical findings include:

        1. **Cataracts — rubella**

        2. **Intracranial calcifications — cytomegalovirus, toxoplasmosis,** or **HIV infection**

        3. **Retinitis — toxoplasmosis** and **rubella**

        4. **Skin lesions — herpesvirus**

IV. **Mental retardation**

    A. Characterized by intellectual-cognitive and social-behavioral functioning significantly below norm for age level; important early clue is delay in normal ability by age 2 years to express two-word phrases (subject-verb; for example, "me eat")

# DEVELOPMENTAL DISORDERS

B. Must be distinguished from:

1. Deafness or hearing impairment

2. **Developmental language disorders** — **learning disabilities** impairing only **specific cognitive tasks** such as **reading (developmental dyslexia)** or **speaking (developmental dysphasia)**

3. **Autism (pervasive developmental disorder)** — syndrome of **abnormal interpersonal relationships** (relates to people as objects), **bizarre mannerisms** (stereotypy such as rocking) and compulsions, and **aberrant language** (echolalia and perseveration)

C. Common causes include:

1. Congenital hypothyroidism (**cretinism**) — **triad of large tongue, abdominal distention, and constipation**, plus lethargy, floppiness, and prolonged neonatal jaundice; diagnosis by low blood thyroxine levels; profound mental retardation, if treatment with thyroid supplementation is delayed beyond age 3 months

2. **Phenylketonuria (PKU)** — normal-appearing neonate with failure to gain weight and feeding difficulties; diagnosis by elevated phenylalanine level in heel-stick blood sample (screening done after milk feeding by law in all states) or by positive urine ferric chloride test; seizures, growth failure, and profound mental retardation if treatment with low phenylalanine diet is delayed; autosomal recessive inheritance.

3. **Down syndrome** — trisomy chromosome 21 malformation syndrome; characterized by varying degrees of mental retardation (mild to profound), hypotonia, oblique palpebral fissures, **epicanthal folds, iris Brushfield spots** (light speckling), **bilateral transverse palmar (simian) creases**, and cardiac, gastrointestinal (**tracheoesophageal fistula or duodenal atresia**), and upper cervical vertebral anomalies; progressive personality changes and intellectual deterioration (dementia) beginning in third decade are associated with pathologic changes of Alzheimer's disease (neurofibrillary tangles and neuritic plaques)

4. **Fragile-X syndrome** — moderate to profound mental retardation, **large ears**, long face with **prominent jaw** and forehead, and **macroorchidism**; X-linked disorder affecting males, but some females carriers can be retarded; chromosome studies using special folate-deficient medium show excessive breaks in long arm of X-chromosome; genetic defect consists of expansion (increased length) of unstable trinucleotide (CGG) repeat sequence in gene in Xq27.3 region

V. **Neurogenetic disorders** — various inherited metabolic diseases result in progressive neurologic disability; definitive diagnosis requires identification of specific enzymatic abnormality

A. **Tay-Sachs disease**

1. Progressive degenerative neurologic disease resulting from **deficiency of enzyme hexosaminidase A** (gene on chromosome 15) resulting in storage of **GM$_2$-ganglioside** in **lysosomes** of neurons

2. Initial clinical symptom of **hyperacusis** (exaggerated startle to sound) in first months of life; followed by **delayed motor development** (poor sitting ability at 6 months), **hypotonia**, loss of head control, and apparent blindness; funduscopic examination reveals **macular cherry red spot**; infant becomes relatively unresponsive, has decerebrate posturing and **opisthotonos** (spasmodic posturing in which spine and extremities are bent backward), and develops frequent myoclonic jerks and convulsions by age 12 to 18 months; head enlargement (**macrocrania**) becomes apparent after age 15 months; death generally occurs by age 2 years from intercurrent infection (usually pneumonia)

---

**SELECTED NEUROGENETIC DISORDERS**

Adrenoleukodystrophy
 (peroxisomal enzyme defect)
Ceroid lipofuscinosis
G$_{M1}$-Gangliosidosis
 ($\beta$-galactosidase A, B, C deficiency)
Krabbe's disease
 (galactocerebroside $\beta$-galactosidase deficiency)
Metachromatic leukodystrophy
 (arylsulfatase A deficiency)
Niemann-Pick disease
 (sphingomyelinase deficiency)
Pompe's disease
 (acid maltase deficiency)
Tay-Sachs disease
 (hexosaminidase A deficiency)

---

B. **Adrenoleukodystrophy**

1. Inherited **demyelinating disease** of **males** resulting from defective gene on **X chromosome** (Xq28 region); enzymatic defect in **peroxisomes** involving **$\beta$-oxidation of very long chain fatty acids** (unbranched saturated fatty acids of 24 to 30 carbon chain length)

2. Initial presentation in childhood with **behavioral abnormalities** (poor school performance and inattentiveness); progresses to **dementia, visual loss, deafness, spasticity,** and unresponsiveness; **brown skin pigmentation** is evident early in disease process, but **adrenal insufficiency** is usually insidious in onset

3. Radiologic imaging studies demonstrate **extensive demyelination**, greater in posterior cerebral hemispheres

VI. **Hydrocephalus** and **spinal dysraphism**

A. **Hydrocephalus** — **ventricular enlargement** due to obstruction of normal cerebrospinal fluid circulation

## DEVELOPMENTAL DISORDERS

       1. **Noncommunicating hydrocephalus — obstruction within ventricular cavities** (commonly at foramen of Monro, aqueduct of Sylvius, or fourth ventricular foramina of Luschka and Magendie) preventing cerebrospinal flow into subarachnoid space

       2. **Communicating hydrocephalus** — cerebrospinal fluid flows from ventricular cavities into subarachnoid space; **obstruction within subarachnoid space** or at absorptive sites in arachnoid villi prevents normal circulation

   B. **Dysraphism — abnormal neural tube development** resulting in **myelomeningocele**; associated with brain maldevelopment resulting in aqueductal stenosis and brain stem and cerebellar tonsils displacement through foramen magnum into spinal canal (**Chiari malformation**)

   C. **Dandy-Walker malformation — midline cerebellar agenesis** with associated **fourth ventricular cyst**

VII. **Anencephaly** — absent calvarium with **remnant malformed gliovascular tissue** due to abnormal rostral neural tube development; prenatal diagnosis by finding **elevated α-fetoprotein levels** in amniotic fluid and maternal blood

VIII. **Attention deficit hyperactivity disorder (ADHD**; developmental hyperactivity)

   A. Described as "always on the go," impulsive, distractible, immature, clumsy, inattentive

   B. Boys more frequently affected than girls.

   C. "Paradoxical reaction" to certain drugs — barbiturates (sedative in normal individuals) markedly increase hyperactivity, whereas stimulants such as amphetamines, methylphenidate, or pemoline reduce hyperactivity and increase attention

IX. **Developmental specific learning disabilities** — normally intelligent child with greater than expected difficulty in acquiring one or more basic academic skills of reading, writing. arithmetic, or spelling; specific syndromes include developmental **dyslexia (specific reading difficulty)**; developmental dyscalculia (specific arithmetic difficulty), developmental dysgraphia (specific writing difficulty), developmental apraxia (extreme clumsiness), and developmental dysphasia (specific spoken language difficulty)

# SELECTED ADDITIONAL READING

Aicardi J: *Epilepsy in Children*. ed 2. New York, Raven Press, 1994.

Arenberg IK, Smith DB (eds): Diagnostic neurotology. *Neurol Clin* 1990; 8(2):199-481.

Baloh RW, Honrubia V: *Clinical Neurophysiology of the Vestibular System*. ed 2. Philadelphia, FA Davis Co, 1990.

Brodal A: Self-observations and neuro-anatomical considerations after a stroke. *Brain* 1973; 96:675-694.

Brumback RA, Leech RW: *Color Atlas of Muscle Histochemistry*. Littleton, Mass, PSG Publishing Co, 1984.

Burger PC, Scheithauer BW, Vogel FS: *Surgical Pathology of the Nervous System and Its Coverings*. ed 3. New York, Churchill Livingstone Inc, 1991.

Byrne TN, Waxman SG: *Spinal Cord Compression: Diagnosis and Principles of Management*. Philadelphia, FA Davis Co, 1990.

Calne DB (ed): *Neurodegenerative Diseases*. Philadelphia, W.B. Saunders Co, 1994.

Coffey CE, Cummings JL (eds): *Textbook of Geriatric Neuropsychiatry*. Washington, DC, American Psychiatric Press, 1994.

Coleman RM: *Wide Awake at 3:00 A.M.* New York, WH Freeman and Co, 1986.

DeMyer W: *Technique of the Neurologic Examination: A Programmed Text*. ed 4. New York, McGraw-Hill Book Co, 1992.

Diagnostic Classification Steering Committee, Thorpy MJ, Chairman. *International Classification of Sleep Disorders: Diagnostic and Coding Manual*. Rochester, MN, American Sleep Disorders Association, 1990.

Duckett S (ed): *The Pathology of the Aging Human Nervous System*. Philadelphia, Lea & Febiger, 1991.

Duckett S (ed): *Pediatric Neuropathology*. Baltimore, Williams & Wilkins, 1995.

Dyck PJ, Thomas PK, Griffin JW, Low PA, Poduslo JF (eds): *Peripheral Neuropathy*. ed 3. Philadelphia, WB Saunders Co, 1993.

Engel AG, Frazini-Armstrong C (eds): *Myology*. ed 2. New York, McGraw-Hill Book Co, 1994.

Engel J Jr: *Seizures and Epilepsy*. Philadelphia, FA Davis Co, 1989.

Freeman JM, Vining EPG, Pillas DJ: *Seizures and Epilepsy in Childhood: A Guide for Parents*. Baltimore, Johns Hopkins University Press, 1990.

Friede RL: *Developmental Neuropathology*. 2nd rev. and expanded ed. Berlin, Springer-Verlag, 1989.

Hopkins A (ed): *Headache: Problems in Diagnosis and Management*. Philadelphia, WB Saunders Co, 1988.

# SELECTED ADDITIONAL READING

Jankovic J, Tolosa E (eds): *Parkinson's Disease and Movement Disorders*. Baltimore, Urban & Schwarzenberg, 1988.

Katzman R, Rowe JW: *Principles of Geriatric Neurology*. Philadelphia, FA Davis Co, 1992.

Leigh RJ, Zee DS: *The Neurology of Eye Movements*. ed 2. Philadelphia, FA Davis Co, 1991.

Mayo Clinic and Mayo Foundation: *Clinical Examinations in Neurology*. Chicago, Mosby-Year Book, 1990.

Midroni G, Bilbao JM: *Biopsy Diagnosis of Peripheral Neuropathy*. Boston, Butterworth-Heinemann, 1995.

More NL, Robins PO: *The 36-Hour Day: A Family Guide to Caring for Persons with Alzheimer's Disease*. Baltimore, Johns Hopkins Press, 1982.

Nyhan WL, Sakati NO: *Diagnostic Recognition of Genetic Disease*. Philadelphia, Lea & Febiger, 1987.

Plum F, Posner J: *The Diagnosis of Stupor and Coma*. ed 3. Philadelphia, FA Davis Co, 1980.

Rapin I: *Children with Brain Dysfunction: Neurology, Cognition, Language, and Behavior*. New York, Raven Press, 1982.

Resor SR Jr, Kutt H (eds): *The Medical Treatment of Epilepsy*. New York, Marcel Dekker Inc, 1992.

Rodnitzky RL: *Van Allen's Pictorial Manual of Neurologic Tests*. Chicago, Mosby-Year Book, 1988.

Ropper AH, Kennedy SK, Zervas N: *Neurobiological and Neurosurgical Intensive Care*. Baltimore, University Park Press, 1983.

Ropper AH, Wijdicks EFM, Truax BT: *Guillain-Barré Syndrome*. Philadelphia, FA Davis Co, 1991.

Russell DS, Rubinstein LJ: *Pathology of Tumors of the Nervous System*. ed 5. Baltimore, Williams & Wilkins, 1989.

Schaumburg HH, Berger AR, Thomas PK: *Disorders of Peripheral Nerves*. ed 2. Philadelphia, FA Davis Co, 1992.

Scheinberg LC (ed): *Multiple Sclerosis. A Guide for Patients and Their Families*. New York, Raven Press, 1983.

Sterns RH, Riggs JE, Schochet SS Jr: Osmotic demyelination syndromes following correction of hyponatremia. *N Engl J Med* 1986; 314:1535-1542.

Task Force for the Determination of Brain Death in Children: Guidelines for the determination of brain death in children. *Ann Neurol* 1987; 22:616-617.

Terry RD (ed): *Aging and the Brain*. New York, Raven Press, 1988.

Victor M, Adams RD, Collins GH: *The Wernicke-Korsakoff Syndrome*. ed 2. Philadelphia, FA Davis Co, 1989.

Volpe J: *Neurology of the Newborn*. ed 3. Philadelphia, WB Saunders Co, 1995.

Wood M, Anderson M: *Neurological Infections*. London, WB Saunders Co, 1988.

# SELF-ASSESSMENT EXAMINATION

1. *Normal* retinal variations visible with the ophthalmoscope include which of the following:
   A. Flame hemorrhages
   B. Cotton wool spots
   C. Optic nerve drusen
   D. Papillitis
   E. Papilledema

2. During the examination of a 44 year old woman with a facial asymmetry, touching the cornea of either eye results in blink in only the right eye, although the patient indicates feeling the touch in both eyes. The most likely lesion is:
   A. Left abducens nerve palsy
   B. Right trochlear nerve palsy
   C. Left trigeminal nerve palsy
   D. Left facial nerve palsy
   E. Right oculomotor nerve palsy

3. The Babinski reflex is:
   A. Dorsiflexion of the big toe and fanning of the other toes following plantar stimulation
   B. Puckering of the lips in response to gentle tapping of the upper lip
   C. Brief visible muscle twitches following needle insertion
   D. Sudden flexion of the hyperextended wrist (flapping motion)
   E. Falling from a standing position following eye closure

4. A 24 year old woman presents after awakening in the morning with slight headache, a generalized sense of fatigue, and visual loss. Examination reveals only a minimal light perception. Funduscopic exam is normal. The most likely diagnosis is:
   A. Papillitis
   B. Papilledema
   C. Retrobulbar neuritis
   D. Uhthoff phenomenon
   E. Lhermitte's symptom

5. *Normal* cerebrospinal fluid values are:
   A. Opening pressure > 200 mm of water
   B. Glucose level less than 40 mg/dL
   C. Up to five lymphocytes per cubic millimeter
   D. Protein level greater than 65 mg/dL
   E. Oligoclonal bands

6. The triad of miosis, ptosis, and anhidrosis characterizes:
   A. Partial oculomotor nerve palsy
   B. Trochlear nerve palsy
   C. Internuclear ophthalmoplegia
   D. Argyll Robertson pupil
   E. Horner's syndrome

7. The Romberg test is useful in evaluation of:
   A. Auditory acuity
   B. Pupillary reflexes
   C. Loss of proprioception
   D. Choreoathetosis
   E. Muscle strength

8. Signs of right hypoglossal nerve palsy include:
   A. Paralysis of head turning to right side
   B. Tongue deviation to right side
   C. Nystagmus with slow component to left side
   D. Right side anhidrosis, miosis, and ptosis
   E. Jaw deviation to left side

9. Optic atrophy is characterized by:
   A. Normal visual acuity
   B. Pale sharply marginated optic disk
   C. Blurring of optic disk margins
   D. Small refractile bodies elevating optic nerve
   E. Venous enlargement

10. A 45 year old man presents with a history of galactorrhea and loss of libido. At another clinic he had an MRI study of the brain that showed a large pituitary adenoma which had extended upward through the diaphragma sellae and was impinging on

the middle of the optic chiasm. The visual field finding associated with such a lesion would be:
A. Hemianopsia
B. Bitemporal hemianopsia
C. Homonymous quadrantanopsia
D. Unilateral blindness
E. Binasal hemianopsia

11. A 60 year old man has a past history of a transient ischemic attack. Such a transient ischemic attack would have been characterized by:
A. Rapidly developing neurologic deficit
B. Persistence for more than 24 hours
C. Pale (bland or white) infarction
D. Lipohyalinosis of small penetrating arteries
E. Severe headache, photophobia, and stiff neck

12. A 55 year old man with known atherosclerotic cardiovascular disease has a stroke that produces the locked-in syndrome. What of the following is the most likely etiology:
A. Middle cerebral artery stroke syndrome
B. Anterior cerebral artery stroke syndrome
C. Posterior cerebral artery stroke syndrome
D. Wallenberg syndrome
E. Basilar artery occlusion

13. A 42 year old physician has sudden onset of severe headache, photophobia, and stiff neck after just scoring a double bogey on the 15th hole of a charity golf tournament. His golfing partner, a psychiatrist, notes no obvious neurologic deficit. The most likely diagnosis is:
A. Thromboembolic stroke
B. Ruptured berry aneurysm
C. Temporal arteritis
D. Psychogenic headache
E. Migraine headache

14. A 48 year old man collapses from a cardiac arrest while walking through the local shopping mall. After about 12 minutes the emergency medical team is able to obtain a stable pulse and blood pressure. Three weeks later, in the hospital, he remains comatose, but breathes without assistance and has occasional reflex limb movements when stimulated. The most likely explanation for his neurologic condition is:
A. Middle cerebral artery stroke syndrome
B. Laminar cortical necrosis
C. Wallenberg syndrome
D. Lacunar stroke
E. Global aphasia

15. Two years following her left hemisphere stroke, a 58 year old woman is brought to the clinic by her son for neurologic evaluation. During the examination, the woman speaks in only short poorly-articulated phrases although she understands all the examiner's verbal instructions. Her handwriting is messy and she cannot repeat any spoken words. She also has a mild right hemiparesis. The most likely diagnosis is:
A. Schizophrenia
B. Broca's aphasia
C. Malingering
D. Wernicke's aphasia
E. Wernicke-Korsakoff psychosis

16. A 25 year old man has had temporal lobe epilepsy for the past 9 years. He also has evidence of behavioral problems associated with depression. Which of the following anticonvulsants would be most likely to control both his behavioral problems and the epilepsy:
A. Carbamazepine (Tegretol)
B. Phenytoin (Dilantin)
C. Lamotrigine (Lamictal)
D. Phenobarbital
E. Ethosuximide (Zarontin)

17. A 6 month old infant has seizures with an EEG pattern of hypsarrhythmia. The most likely epilepsy syndrome to explain this problem would be:
A. West syndrome (infantile spasms)
B. Petit mal epilepsy
C. Benign centrotemporal epilepsy
D. Benign (simple) febrile seizures
E. Temporal lobe epilepsy

18. A 30 year old woman has a history of epilepsy for about the past 12 years. She only rarely has a generalized convulsion. More often she has periods in which she becomes confused. She also has frequent complaint of auras consisting of déjà vu, epigastric sensations, and occasional unpleasant

visual hallucinations. The most likely type of epilepsy is:
A. Lennox-Gastaut syndrome
B. Petit mal epilepsy
C. Benign centrotemporal epilepsy
D. Temporal lobe epilepsy
E. Focal motor seizures

19. The second grade teacher of a 8 year old girl has sent notes home to the parents indicating that the child seems to be daydreaming a lot. The teacher has noted that the child has momentary lapses in which she is unresponsive and occasionally has some eyelid fluttering. The child's physician has indicated that the problem is a form of epilepsy. The most likely diagnosis is:
A. Myoclonic seizure
B. Astatic seizure
C. Absence seizure
D. Interictal seizure
E. Simple partial seizure

20. A 6 year old child has frequent epileptic spells consisting of a blank stare and eyelid fluttering. An EEG shows frequent 3 Hz spike-wave discharges activated by hyperventilation. The first drug of choice for treatment of this child's epilepsy would be:
A. Carbamazepine (Tegretol)
B. Phenytoin (Dilantin)
C. Phenobarbital
D. Primidone (Mysoline)
E. Ethosuximide (Zarontin)

21. Childhood colic, motion sickness, or episodic abdominal pain often precede which later disorder:
A. Migraine
B. Ménière's disease
C. Temporal arteritis
D. Trigeminal neuralgia
E. Pseudotumor cerebri

22. A 45 year old man has headaches that are usually associated with a partial Horner's syndrome. The most likely diagnosis:
A. Tension headache
B. Cluster headache
C. Trigeminal neuralgia
D. Post-lumbar puncture headache
E. Pseudotumor cerebri

23. A 55 year old man complains of frequent episodes of brief paroxysmal lancinating face pain. The pain can be triggered by simply touching the skin adjacent to his right nostril. The most likely diagnosis is:
A. Meralgia paresthetica
B. Trigeminal neuralgia
C. Wallenberg syndrome
D. Bell's palsy
E. Lambert-Eaton syndrome

24. Which disorder is most often associated with papilledema:
A. Tension headache
B. Cluster headache
C. Trigeminal neuralgia
D. Post-lumbar puncture headache
E. Pseudotumor cerebri

25. Temporal arteritis is characterized by all the following EXCEPT:
A. Polymyalgia rheumatica
B. Visual disturbances
C. Granulomatous inflammation
D. Response to corticosteroids
E. High incidence in obese women of childbearing age

26. A 35 year old man presents to his physician with complaints of increasing headache and lethargy. Examination reveals erythematous, hyperkeratotic hands and feet along with white bands in his fingernails. The likely diagnosis is:
A. Mercury poisoning
B. Ethylene glycol poisoning
C. Thiamine deficiency
D. Uremia
E. Arsenic poisoning

27. A college student presents to the hospital with relatively acute onset of fever, severe abdominal pain, confusion, and evidence of a rapidly progressive peripheral and cranial neuropathy, several hours after a wild party at a local bar celebrating his twenty-first birthday. Cerebrospinal

fluid glucose and protein levels are normal. The most likely diagnosis is:
A. Porphyria
B. Wernicke-Korsakoff psychosis
C. Alcohol withdrawal syndrome
D. Guillain-Barré syndrome
E. Methanol poisoning

28. A 4 year old child living in a poor neighborhood presents to the hospital with irritability, lethargy, and ataxia. Mother states that for the past several months, the child has been chronically constipated and complained of abdominal pain. Laboratory studies reveal anemia and basophilic stippling of red blood cells. The most likely diagnosis is:
A. Methanol poisoning
B. Cocaine poisoning
C. Lead poisoning
D. Mercury poisoning
E. Vitamin A overdose

29. An infant born at home without medical care has ABO (blood group) incompatability and develops severe hyperbilirubinemia during the neonatal period. Expected neurologic sequelae would be:
A. Kernicterus
B. Hepatic encephalopathy
C. Porphyria
D. Wilson's disease
E. Wernicke's encephalopathy

30. A 48 year old chronic alcoholic man is admitted to the hospital for treatment of a severe scalp laceration. On the first hospital day the nurses noted his tremulousness, and on the second hospital day he has 3 brief generalized convulsions. The most likely diagnosis is:
A. Petit mal epilepsy
B. Temporal lobe epilepsy
C. Alcohol withdrawal seizures
D. Alcohol-induced hypoglycemic seizures
E. Seizures resulting from cortical contusion

31. After 72 hours of alcohol abstinence, confusion, agitation, tremor, autonomic nervous system hyperactivity, and hallucinations become apparent in a 45 year old known chronic alcoholic man. The most likely diagnosis is:
A. Wernicke's encephalopathy
B. Korsakoff's psychosis
C. Chronic auditory hallucinosis
D. Delirium tremens
E. Chronic subdural hematoma

32. A 53 year old woman with a long history of binge alcohol abuse presents to the emergency department with confusion, ataxia, and external ophthalmoplegia consistent with Wernicke's encephalopathy. Emergency treatment should consist of administration of parenteral:
A. Thiamine
B. Niacin
C. Pyridoxine
D. Vitamin $B_{12}$
E. Phenytoin

33. A lethargic 35 year old skid row alcoholic man is brought to the emergency room by ambulance in coma following several seizures. Laboratory studies indicate severe acidosis, elevated blood urea nitrogen (BUN) level, and cerebrospinal fluid pleocytosis (100 lymphocytes/ per cubic millimeter). Urinalysis shows numerous oxalate crystals. The most likely diagnosis is:
A. Diabetic ketoacidosis
B. Subdural hematoma
C. Ethylene glycol poisoning
D. Pneumococcal meningitis
E. Wood alcohol poisoning

34. A 20 year old woman is brought to the emergency room in coma with findings of slow shallow breathing, pinpoint pupils, bradycardia, and hypothermia. The most appropriate emergency treatment would be administration of:
A. Thiamine
B. Clonidine (Catapres)
C. Diazepam (Valium)
D. Naloxone (Narcan)
E. 100% oxygen

35. A difficult 9 year old girl with cystic fibrosis often does not follow prescribed medical treatment. The mother is aware of frequent diarrhea. Over the past 18 months, the child has developed a progressive peripheral neuropathy and ataxia that mimics

spinocerebellar degeneration. There is no family history of neurologic disease. The most likely explanation is:
A. Hypervitaminosis A
B. Vitamin E deficiency
C. Thiamine deficiency
D. Vitamin K deficiency
E. Hypervitaminosis D

36. An elderly man has had recent onset of symptoms consisting of unsteady gait, dementia, and urinary incontinence. Following a lumbar puncture, these symptoms improve. The most likely diagnosis is:
A. Depressive pseudodementia
B. Alzheimer's disease
C. Normal pressure hydrocephalus
D. Lacunar state
E. Binswanger's disease

37. A 40 year old woman with Down syndrome has recently lost many of her usual abilities of self-care and language. Histopathologic examination of her brain would most likely show characteristics of:
A. Alzheimer's disease
B. Pick's disease
C. Creutzfeldt-Jakob disease
D. Wilson's disease
E. Parkinson's disease

38. An 18 year old boy has developed progressive ataxia, areflexia, spasticity, pes cavus foot deformity, and cardiomyopathy during the previous four years. This most likely diagnosis is:
A. Friedreich's ataxia
B. Progressive multifocal leukoencephalopathy
C. Creutzfeldt-Jakob disease
D. Ataxia-telangiectasia
E. Wilson's disease

39. The infectious agent of Creutzfeldt-Jakob disease is identified as a
A. Pick body
B. Lewy body
C. Negri body
D. Neuritic plaque
E. Prion

40. A 48 year old woman presents with rapidly progressive dementia, ataxia, and myoclonus. The most likely diagnosis is:
A. Alzheimer's disease
B. Pick's disease
C. Creutzfeldt-Jakob disease
D. Wilson's disease
E. Parkinson's disease

41. The 55 year old rotund head chef at a four-star restaurant has a history of heart disease that required coronary artery bypass surgery. One day while at work, he suddenly collapses. When the emergency medical team arrives, exam reveals 4 mm diameter symmetric unreactive pupils (midposition non-reactive pupils). The most likely site for the brain lesion producing this clinical picture is:
A. Right frontal lobe
B. Bilateral occipital poles
C. Medulla
D. Pons
E. Midbrain

42. A 52 year old man with a history of chronic hypertension presents to the emergency department with sudden onset of vomiting, dizziness, ataxia, and leg weakness. His blood pressure is 240/110 and his pulse is 55 beats per minute. In the few minutes following his initial presentation to the emergency department, he becomes comatose, develops ataxic breathing, and then dies. His clinical symptomatology is most readily explainable as the result of:
A. Transtentorial uncal herniation
B. Central rostral-caudal herniation
C. Subfalcial herniation
D. Cerebellar tonsillar herniation

43. Following resuscitation for a cardiac arrest, a 65 year old man displays no responsiveness to stimuli, occasional spontaneous eye opening, and an EEG pattern consistent with a nearly normal sleep-wake cycle. This patient would be considered to have:
A. Coma
B. Locked-in syndrome
C. Persistent vegetative state
D. Delirium
E. Normal state of consciousness

44. A 50 year old pedestrian struck by an automobile has progressive signs consisting of initial pupillary dilation followed by loss of extraocular movements, contralateral hemiparesis, coma, and medullary dysfunction. The most likely explanation is:
    A. Transtentorial uncal herniation
    B. Central rostral-caudal herniation
    C. Subfalcial herniation
    D. Cerebellar tonsillar herniation

45. Brain death requires all the following criteria to be met EXCEPT:
    A. Established coma-causing cerebral lesion
    B. Apnea
    C. Absent spinal reflexes
    D. Absent brain stem reflexes
    E. Absence of toxins or metabolic abnormalities

46. A 24 year old woman was involved in an automobile accident, in which her car ran off a deserted mountain road and struck a tree. When she was discovered by a passerby some time later, the woman was alert but complaining of headache and had bruises on her forehead and face. By the time she arrives at a hospital emergency room (about an hour away), the personnel note she is lethargic but arousable, and has no obvious neurologic deficits on examination. Over the next hour she becomes progressively less arousable, her left pupil appears larger than her right pupil, and her right arm and leg seem weak. What is the most likely diagnosis?
    A. Chronic subdural hematoma
    B. Carotid-cavernous fistula
    C. Epidural hematoma
    D. Cerebral concussion
    E. Diffuse axonal injury

47. A 12 year old boy is brought to the emergency room by his mother after a fall from his bicycle in which he struck his head. He apparently was momentarily unconscious, but subsequently he seemed perfectly fine except for complaints of a slight headache. Examination reveals only a small bruise on his forehead and neurologic exam is normal. The most likely diagnosis is:
    A. Chronic subdural hematoma
    B. Carotid-cavernous fistula
    C. Epidural hematoma
    D. Cerebral concussion
    E. Diffuse axonal injury

48. The Brown-Séquard syndrome is characterized by:
    A. Ipsilateral spasticity and proprioceptive loss and contralateral loss of pain and temperature sensation
    B. Greater weakness in arms than in legs, patchy sensory loss, and urinary retention
    C. Bilateral spasticity and loss of pain and temperature sensation with preservation of proprioception
    D. Bilateral flaccid paralysis, anesthesia, areflexia, and bladder and sphincter dysfunction
    E. Bilateral loss of proprioception

49. A 73 year old woman is brought to the hospital by her son because of recent onset of confusion. The son indicates that his mother had been complaining of headaches for several weeks, ever since she was "roughed up" by hooligans trying to steal her purse. On examination, she appears drowsy, is unable to identify her surroundings, does not know the date or her son's name, and is weak on her left side. What is the most likely diagnosis?
    A. Chronic subdural hematoma
    B. Carotid cavernous fistula
    C. Epidural hematoma
    D. Cerebral concussion
    E. Diffuse axonal injury

50. The most important factor in post-traumatic epilepsy is:
    A. Lacunar infarcts
    B. Orbital frontal plaque jaune
    C. Duret hemorrhages
    D. Arachnoidal fibrosis
    E. Ventricular dilation

51. A family presents to the neuromuscular clinic with a three generation history of distal muscle atrophy, pes cavus foot deformity, and sensory loss. The most likely diagnosis is:
    A. Amyotrophic lateral sclerosis
    B. Charcot-Marie-Tooth disease
    C. Guillain-Barré syndrome
    D. Tardy ulnar palsy
    E. Meralgia paresthetica

## SELF-ASSESSMENT EXAMINATION

52. Following a brief flu-like illness, a 24 year old man presents with subacute onset over 5 days of symmetric motor weakness, greater in the arms than in the legs. Lumbar puncture reveals moderately elevated CSF protein with few cells (albuminocytologic dissociation). The most likely diagnosis is:
    A. Guillain-Barré syndrome
    B. Porphyria
    C. Vitamin $B_{12}$ deficiency
    D. Alcoholic polyneuropathy
    E. Diabetic amyotrophy

53. A 24 year old woman complains of increasing weakness throughout the day, despite feeling strong upon awakening in the morning. In the evenings, she also notes drooping of one eyelid and occasional double vision. The most likely diagnosis is:
    A. Polymyositis
    B. Duchenne dystrophy
    C. Myotonic dystrophy
    D. Malignant hyperthermia
    E. Myasthenia gravis

54. A 55 year old moderately obese woman complains of burning paresthesias and loss of sensation on her lateral thigh. The most likely diagnosis is:
    A. Meralgia paresthetica
    B. Trigeminal neuralgia
    C. Wallenberg syndrome
    D. Bell's palsy
    E. Lambert-Eaton syndrome

55. A 14 year old previously health boy experiences a mild gastroenteritis followed by the rapid onset of flaccid weakness in the right arm and left leg. The most likely diagnosis is:
    A. Toxoplasmosis
    B. Tabes dorsalis
    C. Mucormycosis
    D. Cysticercosis
    E. Poliomyelitis

56. Horner's syndrome often accompanies which of the following conditions:
    A. Charcot-Marie-Tooth disease
    B. Meralgia paresthetica
    C. Erb-Duchenne palsy
    D. Klumpke-Dejerine palsy
    E. Bell's palsy

57. A frail 60 year old widow who lives alone and has no friends in the community presents to her physician with a complaint of feeling chronically tired. On exam, she is noted to have loss of posterior column sensation, a positive Romberg test, spasticity, and bilateral Babinski reflexes. The most likely diagnosis is:
    A. Guillain-Barré syndrome
    B. Porphyria
    C. Vitamin $B_{12}$ deficiency
    D. Alcoholic polyneuropathy
    E. Diabetic amyotrophy

58. Foot drop would be expected with:
    A. Femoral nerve palsy
    B. Peroneal nerve palsy
    C. Erb-Duchenne palsy
    D. Klumpke-Dejerine palsy
    E. Meralgia paresthetica

59. A 55 year old woman with 70 pack-year history of smoking presents with complaints of generalized weakness. One month previously she had been diagnosed with small cell (oat cell) lung cancer. Repetitive nerve stimulation studies performed by the clinical neurophysiology laboratory reveal an incrementing response of the muscle action potentials. The most likely diagnosis is:
    A. Lambert-Eaton syndrome
    B. Botulism
    C. Malignant hyperthermia
    D. Myotonic dystrophy
    E. Myasthenia gravis

60. The disorder associated with dystrophin deficiency is:
    A. Myotonic dystrophy
    B. Polymyositis
    C. Myasthenia gravis
    D. Duchenne dystrophy
    E. Type 2 muscle fiber atrophy

61. A 59 year old hypertensive man presents to the emergency department with the acute onset of severe vertigo, nausea, vomiting, nystagmus, and difficulty

swallowing. Exam shows a moderately severe gait ataxia, right arm dysmetria, loss of the corneal reflex on the right with reduced sensibility on the right side of the face, a right Horner's syndrome, and diminished sensation to pinprick on the left arm, trunk, and leg. He has no demonstrable limb weakness. The most likely diagnosis is:
A. Ménière's disease
B. Wallenberg syndrome
C. Acoustic neuroma
D. Basilar artery migraine
E. Vestibular neuronitis

62. The illusion that stationary objects are moving back and forth is:
A. Oscillopsia
B. Dysequilibrium
C. Dizziness
D. Nystagmus
E. Vertigo

63. A 30 year old man complains of recurrent sudden attacks of vertigo associated with tinnitus. Audiometry indicates progressive high tone hearing loss. The most likely diagnosis is:
A. Ménière's disease
B. Benign positional vertigo
C. Vestibular neuronitis
D. Motion sickness
E. Basilar artery migraine

64. Acute onset of vertigo and nystagmus associated with viral nasopharyngitis is characteristic of:
A. Ménière's disease
B. Benign positional vertigo
C. Vestibular neuronitis
D. Motion sickness
E. Basilar artery migraine

65. The illusion of rotational movement of self or the environment is termed:
A. Oscillopsia
B. Dysequilibrium
C. Dizziness
D. Nystagmus
E. Vertigo

66. A 30 year old man presents with recent onset of ataxia, dizziness, and headache. Complete blood count indicates polycythemia. Magnetic resonance imaging (MRI) of the brain reveals a cystic cerebellar tumor near the foramen magnum. Computed tomographic (CT) scans of the abdomen identify cysts of the kidney and pancreas. The most likely diagnosis for the brain tumor is:
A. Astrocytoma
B. Meningioma
C. Neurilemoma
D. Colloid cyst of third ventricle
E. Hemangioblastoma

67. A 32 year old man with a fifteen year history of temporal lobe epilepsy has recently had an increasing number of seizures despite the addition of several new medications to his anticonvulsant regimen. CT scan and MRI indicate a partially calcified mass in the anterior temporal lobe. The most likely histologic finding on biopsy of this mass would be:
A. Glioblastoma multiforme
B. Craniopharyngioma
C. Neurilemoma
D. Oligodendroglioma
E. Medulloblastoma

68. Acoustic neuroma (neurilemoma or schwannoma) may be a part of what syndrome:
A. Neurofibromatosis
B. Cushing's disease
C. Von Hippel-Lindau syndrome
D. Acquired immunodeficiency syndrome (AIDS)
E. Ataxia-telangiectasia

69. Which of the following tumors is associated with homozygous deletion of a region on chromosome 13 (region 13q14):
A. Pineal choriocarcinoma
B. Pituitary adenoma
C. Retinoblastoma
D. Meningioma
E. Ependymoma

70. A 38 year old woman has recently noted galactorrhea and amenorrhea. Neurologic exam is

normal except for bitemporal hemianopsia. The most likely diagnosis is:
A. Colloid cyst of third ventricle
B. Pituitary adenoma
C. Pineal dysgerminoma
D. Pituitary apoplexy
E. Trilateral retinoblastoma

71. A 38 year old woman with a history of chronic schizophrenia has been hospitalized for treatment in the state mental facility for the past 22 years. Over these years she has been continuously treated with dopamine-blocking neuroleptic drugs. The attendants have observed that in the recent year she has more repetitive purposeless movements, particularly of the face and mouth. The most likely diagnosis is:
A. Myoclonus
B. Asterixis
C. Parkinsonism
D. Ballismus
E. Tardive dyskinesia

72. A 14 year old girl presents with ataxia and choreoathetosis. The ophthalmologist has identified Kayser-Fleischer rings during an eye examination. The most likely diagnosis is:
A. Huntington's chorea
B. Wilson's disease
C. Gilles de la Tourette syndrome
D. Progressive supranuclear palsy
E. Tardive dyskinesia

73. A young single mother calls for an appointment for her 9 year old son who is now in the third grade. The teacher is complaining that the child disrupts the class because of his frequent facial grimaces, grunting and snorting sounds, and frequently shouted obscenities. The mother says that the child takes no medications and she denies any illicit drug use in the house. A likely explanation for this problem would be:
A. Huntington's chorea
B. Wilson's disease
C. Gilles de la Tourette syndrome
D. Progressive supranuclear palsy
E. Tardive dyskinesia

74. The tremor characteristically associated with Parkinson's syndrome is:
A. Myoclonus
B. Resting tremor
C. Intention tremor
D. Action tremor
E. Benign essential tremor

75. A 39 year old woman is brought to the emergency department for treatment of an attempted suicide in which she took a sublethal dose of the tranquilizer diazepam (Valium). Her husband relates that she has been acting strangely for the past four years, with increased forgetfulness, poor judgment, and irregular jerky limb movements. Apparently her mother had a similar clinical picture and successfully committed suicide at age 37 years. She also has an estranged older brother residing in a state mental hospital for the past 5 years with dementia and a movement disorder. The most likely diagnosis is:
A. Huntington's disease
B. Gilles de la Tourette syndrome
C. Neurofibromatosis type 1
D. Myasthenia gravis
E. Progressive multifocal leukoencephalopathy

76. An organism acquired in the birth canal that frequently causes meningitis in neonates is:
A. *Neisseria meningitidis*
B. *Listeria monocytogenes*
C. *Staphylococcus aureus*
D. *Haemophilus influenzae*
E. *Staphylococcus epidermidis*

77. Examination of a 55 year old retired merchant marine sailor who spent many years working on cargo ships in the Orient reveals marked sensory loss involving mainly proprioception, positive Romberg test, and severely deformed knee joints (Charcot joints). Pinprick exam. He complains bitterly of brief sharp lancinating pains in the legs which can be brought on during the examination by pinprick. The most likely diagnosis is:
A. Toxoplasmosis
B. Tabes dorsalis
C. Mucormycosis
D. Cysticercosis
E. Poliomyelitis

78. The disorder of multiple brain cysts produced by the larval form of the pork tapeworm is:
    A. Neurosyphilis
    B. Lyme disease
    C. Toxoplasmosis
    D. Cysticercosis
    E. Scrub typhus

79. An elderly gentleman presents to the office with the complaint of sharp pains unilaterally in the upper abdomen and back in a band like distribution. Examination reveals reduced sensibility in the same area as the pain along with a vesicular rash. The most likely diagnosis is:
    A. Poliomyelitis
    B. Shingles
    C. Progressive multifocal leukoencephalopathy
    D. Subacute sclerosing panencephalitis
    E. Herpes simplex encephalitis

80. Two unrelated children in the same fifth grade classroom develop meningitis and a petechial skin rash. Both children die within several hours of the onset of the illness and at autopsy the medical examiner notes hemorrhagic infarction of the adrenal glands (Waterhouse-Friderichsen syndrome) in both children. The most likely causative agent is:
    A. *Neisseria meningitidis*
    B. *Listeria monocytogenes*
    C. *Staphylococcus aureus*
    D. *Haemophilus influenzae*
    E. *Staphylococcus epidermidis*

81. A 50 year old man is receiving various drugs for treatment for lymphoma. Over the past several weeks, he has developed a rapidly progressive dementia, associated with signs of ataxia, visual field defects, and spasticity. The most likely explanation is:
    A. Friedreich's ataxia
    B. Progressive multifocal leukoencephalopathy
    C. Creutzfeldt-Jakob disease
    D. Ataxia-telangiectasia
    E. Wilson's disease

82. The vaso-invasive organisms that spread from paranasal sinuses and retro-orbital tissues to produce fatal brain infection in poorly-controlled diabetes mellitus are most likely to be:
    A. *Mycobacterium tuberculosis*
    B. *Cryptococcus neoformans*
    C. Mucormycosis
    D. *Borrelia burgdorferi*
    E. *Toxoplasma gondii*

83. While reviewing the medical record of a 55 year old recent immigrant from a Southeast Asia refugee camp, it is discovered that he has a positive syphilis serology in both his blood and cerebrospinal fluid. The ocular finding that would be useful in making the diagnosis of neurosyphilis in this man is:
    A. Partial oculomotor nerve palsy
    B. Trochlear nerve palsy
    C. Internuclear ophthalmoplegia
    D. Argyll Robertson pupil
    E. Horner's syndrome

84. A 40 year old outdoorsman presents for evaluation of a facial palsy, chronic headache, and leg pain, sensory disturbance, and some weakness in a radicular distribution. Further history indicates that he frequently camps outdoors for weeks at a time throughout the New England area. He admits to having been bitten by ticks, but does not remember when. A likely diagnosis would be:
    A. Neurosyphilis
    B. Lyme disease
    C. Toxoplasmosis
    D. Cysticercosis
    E. Scrub typhus

85. A 26 year old previously healthy woman experiences the subacute onset of behavioral changes, fever, and headache. Following a generalized seizure, she is brought to the emergency department where radiologic imaging studies show hemorrhagic necrosis of inferomedial temporal lobes. The most likely diagnosis is:
    A. Poliomyelitis
    B. Cysticercosis
    C. Progressive multifocal leukoencephalopathy
    D. Subacute sclerosing panencephalitis
    E. Herpes simplex encephalitis

86. The disorder characterized by a fracture of the L5 vertebral neural arch occurring at or before birth is called:
    A. Klippel-Feil anomaly
    B. Spondylolysis
    C. Pott's disease
    D. Ankylosing spondylitis
    E. Syringomyelia

87. A 32 year old woman presents to the hospital because of a severe second-degree burn involving most of the palm of her left hand that occurred when she accidentally laid her hand on the burner of a stove and did not notice that the burner was hot. Exam reveals marked wasting and weakness of all intrinsic hand muscles bilaterally. She has loss of pain and temperature sense over both arms, shoulders, upper trunk and neck. Proprioception is preserved in her arms and hands. She also has bilateral extensor plantar responses (Babinski reflexes). The most likely diagnosis is:
    A. Klippel-Feil anomaly
    B. Spondylolysis
    C. Pott's disease
    D. Ankylosing spondylitis
    E. Syringomyelia

88. The anterior cord syndrome is characterized by:
    A. Ipsilateral spasticity and proprioceptive loss and contralateral loss of pain and temperature sensibility
    B. Greater weakness in arms than in legs, patchy sensory loss, and urinary retention
    C. Bilateral spasticity and loss of pain and temperature sensibility with preservation of proprioception
    D. Bilateral flaccid paralysis, anesthesia, areflexia, and bladder and sphincter dysfunction
    E. Bilateral loss of proprioception

89. Following surgery for an atherosclerotic abdominal aortic aneurysm, a 65 year old man with a 75 pack/year smoking history awakens with paraplegia and loss of pain and temperature sensation extending up to the T10 spinal cord level. Proprioception in his legs is preserved. The operative note from the surgeon indicates a difficult surgical repair of a large aneurysm that involved the origins of the renal arteries (which had to be reimplanted above the dacron bypass graft used in the repair). The most likely diagnosis is:
    A. Syringomyelia
    B. Pott's disease
    C. Anterior spinal artery syndrome
    D. Ankylosing spondylitis
    E. Intramedullary metastasis of lung cancer

90. A 29 year old man complains of stiff back and neck, particularly upon awakening in the morning. Exam shows limited neck motion and limited ability to bend forward at the waist. Spine radiographs show destruction of the sacroiliac joints and early bridging between vertebral bodies in the thoracic and lumbar spine. The most likely diagnosis is:
    A. Klippel-Feil anomaly
    B. Spondylolysis
    C. Pott's disease
    D. Ankylosing spondylitis
    E. Syringomyelia

91. Obstructive sleep apnea occurs during which sleep stage:
    A. Stage I sleep
    B. Stage II sleep
    C. Stage III sleep
    D. Stage IV sleep
    E. REM sleep

92. The symptom of narcolepsy in which there is sudden loss of muscle tone often precipitated by strong emotion (such as laughter) is:
    A. Sleep paralysis
    B. Cataplexy
    C. Hypnogogic hallucinations
    D. Pickwickian syndrome
    E. Pavor nocturnus

93. A 35 year old seemingly-healthy woman undergoing a dexamethasone suppression test shows no evidence of the normal circadian cortisol rhythm. The most likely explanation is:
    A. Somnambulism
    B. Simultaneous bromocriptine administration
    C. Depressive disorder
    D. Lambert-Eaton syndrome
    E. Binswanger's disease

94. Overly aggressive correction of hyponatremia can produce:
    A. Acute disseminated encephalomyelitis
    B. Experimental allergic encephalomyelitis
    C. Central pontine myelinolysis
    D. Acute transverse myelitis
    E. Retrobulbar neuritis

95. A 29 year old woman with multiple sclerosis has the following neurologic findings: with attempts to gaze to the left, the right (adducting) eye does not move past the midline, while the left (abducting) eye moves out but develops nystagmus; she can converge normally. The lesion producing these signs affects:
    A. Medial longitudinal fasciculus
    B. Right optic nerve
    C. Medial lemniscus
    D. Left Edinger—Westphal nucleus
    E. Right oculomotor nerve

96. A retarded teenage boy with large ears, prominent jaw, and macroorchidism would most likely have which disorder:
    A. Phenylketonuria
    B. Down syndrome
    C. Fragile-X syndrome
    D. Cretinism
    E. Adrenoleukodystrophy

97. At 3 months of age, the mother of a previously healthy infant first noted that the child had an exaggerated startle to slight noises in the bedroom. Developmental milestones have been delayed such that now at age 10 months the child still has poor sitting ability. Exam reveals generalized hypotonia. The child does not seem to respond visual cues. The optic fundus shows a macular cherry red spot. The most likely diagnosis is:
    A. Tay-Sachs disease
    B. Adrenoleukodystrophy
    C. Pompe's disease
    D. Autism
    E. Cretinism

98. In a newborn with large tongue, abdominal distention, constipation, lethargy, floppiness, and prolonged neonatal jaundice, the most likely diagnosis is:
    A. Phenylketonuria
    B. Down syndrome
    C. Fragile-X syndrome
    D. Cretinism
    E. Adrenoleukodystrophy

99. Elevated $\alpha$-fetoprotein levels in amniotic fluid are characteristic of which disorder:
    A. Tay-Sachs disease
    B. Adrenoleukodystrophy
    C. Pompe's disease
    D. Phenylketonuria
    E. Anencephaly

100. The disorder associated with abnormal peroxisomal ß-oxidation of very long chain fatty acids is:
    A. Phenylketonuria
    B. Down syndrome
    C. Fragile-X syndrome
    D. Cretinism
    E. Adrenoleukodystrophy

# SELF-ASSESSMENT EXAMINATION

## Answers to Self-Assessment Examination

Chapter 1
1. C
2. D
3. A
4. C
5. C
6. E
7. C
8. B
9. B
10. B

Chapter 2
11. A
12. E
13. B
14. B
15. B

Chapter 3
16. A
17. A
18. D
19. C
20. E

Chapter 4
21. A
22. B
23. B
24. E
25. E

Chapter 5
26. E
27. A
28. C
29. A
30. C
31. D
32. A
33. C
34. D
35. B

Chapter 6
36. C
37. A
38. A
39. E
40. C

Chapter 7
41. E
42. D
43. C
44. A
45. C

Chapter 8
46. C
47. D
48. A
49. A
50. B

Chapter 9
51. B
52. A
53. E
54. A
55. E
56. D
57. C
58. B
59. A
60. D

Chapter 10
61. B
62. A
63. A
64. C
65. E

Chapter 11
66. E
67. D
68. A
69. C
70. B

Chapter 12
71. E
72. B
73. C
74. B
75. A

Chapter 13
76. B
77. B
78. D
79. B
80. A
81. B
82. C
83. D
84. B
85. E

Chapter 14
86. B
87. E
88. C
89. C
90. D

Chapter 15
91. E
92. B
93. C

Chapter 16
94. C
95. A

Chapter 17
96. C
97. A
98. D
99. E
100. E